How T

Stra

Contents

Preface

The main purpose of *How to Win More* is to give you valuable insights into how to improve your long-term return on investment when playing lotto. You shouldn't mistake the title, however. It doesn't mean "*win more often*" ; that will happen only if you buy more tickets. Rather, the title means that your winnings will usually be larger when your numbers do come up.

In addition, reading this book might prevent you from having a great disappointment—perhaps your greatest disappointment ever. Imagine that your dream of a lifetime—winning a jackpot of, let's say, $50,000,000—has come true for you. (That's why you play lotto, isn't it?) However, imagine your disappointment when the lottery company tells you that you have to split that jackpot with more than 10,000 other players, resulting in a share of less than $5,000 for you.

Could this really happen? Yes it could, if, for example, the winning combination was 8, 15, 22, 29, 36, 43. If this combination ever comes up in any number lottery, it will likely produce thousands of very disappointed winners. (The reason for this is explained in Chapter 5.) By the way, 14,697 entries selected this combination in the New York State Lotto of June 7, 1986.

If you are open minded, reading this book will make you an informed lotto player who knows the following:

- The most popular number combinations are also the most foolish. (Many are given in Chapter 5.)

- It is impossible to improve the *odds* of winning any legitimate lottery (for example, by charting numbers).
- The notion of picking so-called hot or cold numbers is nonsense. (You might as well throw your money away.)
- Although nice-looking tickets may be popular, they are also unprofitable.
- Never copy suggested numbers from "Nighthawks," lottery horoscopes ("Zodiac Best Bets"), or "Dr. X's Computer Generated Lotto Picks" (to name only a few supposed consulting experts).
- It is extremely important to avoid picking the winning combinations of past drawings (although each has the same odds of winning as any other combination for future drawings).

Certainly, these maxims, along with other information in this book, will not be confirmed by some agents who profit from lingering lottery myths. Nonetheless, they are only too true.

As you probably know, there are many different number lotteries. To date, Illinois plays a 6/54 (that is, each participant picks six numbers between 1 and 54), California has a 6/51, Canada runs a 6/49, Sweden has a 7/35, and Switzerland plays a 6/45, to name only a few. This book doesn't concentrate on any specific number lottery, for two reasons:

1. All number lotteries are basically the same, so if you understand one, you understand them all.
2. States like to change the types of their lotto games fairly often. For example, California switched from a 6/49 to a 6/53 lotto in June 1990 and to a 6/51 lotto in December 1991, and Arizona changed from a 6/39 to a 6/42 in 1989. The only reason for these changes (and many similar changes in other states) is to keep ticket sales at a high level, which makes the lotto odds

more difficult. As a rule of thumb, the odds for winning a jackpot must be seen in comparison with the population base in the given state. If the odds are too low, there will usually be a jackpot winner. Consequently, jackpots won't get big and ticket sales won't increase. On the other hand, if the odds against winning the jackpot are too high, few grand-prize winners will be presented to the public and ticket sales might taper off. With the right odds, there will sometimes be a string of jackpots with no winners, leading to a huge carryover of money and thus much publicity and greater participation. Moreover, the right odds will also entail enough winning of small prizes for player reinforcement.

Although you will learn about the odds (and much more) for *any* lottery, don't be afraid. The discussion of mathematics will be kept to an absolute minimum, and formulae for computing odds or expectations have been "banned" into a separate unit (Chapter 8).

Acknowledgements

We are indebted to David R. Bellhouse (University of Western Ontario, Canada), Henri Carnal (University of Bern, Switzerland), Persi Diaconis (Harvard University, Boston), Harry Joe (University of British Columbia, Canada), Pekka Kangas (University of Oulu, Finland), Mike Orkin (California State University, Hayward), Hal Stern (Iowa State University), and Donald Ylvisaker (University of California, Los Angeles) for providing many valuable comments on early drafts of this book.

Special thanks go to the following lottery corporations for providing us with valuable information and data: California State Lottery, Colorado Lottery, Connecticut Lottery, Florida State Lottery, Maryland State Lottery, Massachusetts State Lottery, New Zealand Lottery, Ontario Lottery Corporation, Queensland

Lottery, Swiss Lottery, and Toto-Lotto-GmbH, Stuttgart, Germany.

We are very much indebted to Guido Kaufmann (University of Bern, Switzerland) and Thorsten Wagner (University of Bochum, Germany) for making huge data sets of lottery tickets amenable to further investigations, and to our copyeditor, Susan Freese, who prevented a Germanized version of the English language that would have amused our readers. Our thanks also go to everyone at A K Peters for their strong support.

Last but not least, we thank Heiko Zimmermann and Nora Gürtler (University of Karlsruhe, Germany), who provided the figures, and Ingrid Voss, who typed a difficult manuscript with great efficiency.

January, 1998

Norbert Henze, Karlsruhe
Hans Riedwyl, Bern

Chapter 1

The History of Lotteries

The word *lottery* (French: *loterie*; German: *Lotterie*) refers to any arrangement for distributing prizes by chance among purchasers of tickets. It is derived from the word *lot* (Old English: *hlot*; Germanic: *hleut, hlaut*; Dutch: *lot*; Danish: *lod*), which describes an object used in deciding a matter of chance.

Casting lots as a method of selection is probably as old as humankind. Doing so seemingly removes the human element and has even been regarded as a means of following divine will.

The origin of lotto-type number lotteries, such as those in abundance today, was a lottery run in Genoa, Italy, in the seventeenth century. It was derived from elections by lot. However, according to Bellhouse (1991), the generally accepted story that the Genoese-style lottery was invented by Benedetto Gentile in 1610 does not seem fully accurate. Bellhouse (1991, p. 146) states:

> *Some evidence points to the conclusion that during the whole of the seventeenth century there was no council of 90 from which five were elected by lot, but that the lottery was indeed inspired by the Genoese election system,*

and he summarizes on p. 147:

> *The story of Benedetto Gentile, the supposed inventor of the lottery who flourished circa 1620, appears*

> *in Volkmann (1771, pages 839-840), but not in Labat (1730). No source references are given by Volkmann. In view of the other collected information on this lottery, this source is too late to be reliable.*

The whole history of lotteries is characterized by the conflict between greed (on the parts of both the participants in the lottery and the entrepreneurs running it) and ethical objections against gambling. For example, Pope Benedict XIII, whose papacy ran from 1724 to 1730, banned the Genoese lottery in 1725. However, his successor, Pope Clement XII, set up a lottery in Rome only six years later to raise Vatican revenues.

In response to the ethical objections against gambling, the use of lottery profits for supposedly good causes has always been offered to dim opposition from church and other dissenting groups. For example, the first efforts to colonize the United States were financed in the years 1612–1620 by profits from English lotteries earmarked for the Virginia Company. Interestingly, however, at the same time that company officials were attempting to boost lottery sales in England, they were trying to prohibit gambling in Virginia.

At first sight, this logic that encourages gambling in one context but outlaws it in another is somewhat curious. Even so, it has been argued many times. For example, the Puritans of Massachusetts, who were definitely opposed to gambling on moral grounds, discovered the value of the lottery in the early eighteenth century when they needed massive funds for the construction of churches, colleges, and roads. *Pecunia non olet*! (Money doesn't stink!) Other worthy causes that turned to lotteries for raising revenue were the funding of Kings College (later known as Columbia University) in 1746 and the endowments of many other American universities, including Yale, Harvard, Princeton, North Carolina, and Pennsylvania. (For a review of lotteries in colonial America, see Sullivan [1972].)

Due to scandals of mismanagement and even outright fraud—some involving exposing bogus lotteries and selling fictitious tickets—lotteries fell out of favor in the first half of the nineteenth century. By 1862, Missouri and Kentucky were the only U.S. states that still ran lotteries. After the Civil War, there was a short-lived revival, notably in Louisiana, but by 1894, legal lotteries had disappeared from the United States for the next 70 years. Similar developments happened in other countries. In Germany, for example, there were no number lotteries from 1861, when Bavaria banned them, until 1952, when Berlin started Lotto 5/90.

It should be stressed that in contrast to the modern number lotteries—such as the California Super Lotto 6/51, Massachusetts Millions 6/49, Canadian Lotto 6/49, and German Lotto 6/49—the previously mentioned British and American lotteries were carried out using numbered tickets. The winners were chosen by selecting numbers using a wheel of fortune, a procedure that sometimes took several days or even weeks. In most countries, the era of the modern number lotteries, in which the mode of drawing winners is much simpler and faster, started after World War II. State governments recognized that gambling is a universal phenomenon that seems inherent in human nature, making it impossible to prevent people from participating. It was consequently argued that the state would be wise to exploit this phenomenon for the purpose of voluntary taxation. This view of lotto is brought out delightfully in a cynical farce quoted in Morton (1990, p. 16):

Lotto's a taxation
On all fools in the nation
But heaven be praised
It's so easily raised.

(A thorough discussion of the "painless tax" aspect of lotteries is given in Chapter 11 of Clotfelter & Cook [1991].)

By exercising strict control over lotteries, states have been able to raise funds for what they have deemed worthy causes. A typical example is the California Senate Bill No. 91 of January 13,

1971, introduced by Senator Mervyn M. Dymally, which states the following:

> *The purpose of the State Lottery Law is to provide for the organization and administration by the State of California of a lottery the net proceeds of which are to be applied to the support of primary and secondary education in the state.*

Of course, Senator Dymally was followed (and preceded) by numerous other politicians who called for state lotteries and publicly backed them as revenue-raising tools, especially for good purposes such as education (California, Illinois, Florida, Michigan, New York) and senior citizens programs (Pennsylvania, West Virginia). In California, for example, the lottery's contribution to public education is equal to approximately 2% of the state's total funding for the schools. Meanwhile, lotto playing is now an integral part of millions of people's leisure activities all over the world, and state budgets are increasingly dependent on lottery revenues.

	State Lottery	Millions of $ U.S.	Date	Number of tickets
1.	California	118.8	04/17/91	10
2.	Pennsylvania	115.6	04/26/89	14
3.	Wisconsin	111.2	07/07/93	1
4.	Florida	106.5	09/15/90	6
5.	Nebraska/Indiana	101.0	11/30/94	2
6.	New York	90.0	01/26/94	9
7.	Florida	89.7	10/26/91	6
8.	Dist. of Columbia	89.5	12/22/93	2
9.	Florida	86.0	03/20/93	3
10.	Texas	77.1	03/16/94	5

Table 1.1 Top 10 U.S. jackpots (as of February 1997) and number of winning tickets for each.

To enhance the dream of obtaining instant wealth and to capture the imaginations and money of millions of people weekly (or even more frequently), many state lotteries have created jackpots, which are that part of the total prize money to be split among the players who have matched all winning numbers. Since a jackpot rolls over to the next drawing when there is no grand-prize winner, there are occasionally huge jackpots, sometimes involving millions. When this happens, lottomania is rampant. Table 1.1 shows the top 10 U.S. jackpots (as of February 1997) together with the number of winning tickets for each jackpot.

Not only Americans are susceptible to lottomania; it's a worldwide phenomenon. Figure 1.1 shows the weekly sales for the German Lotto 6/49 (in millions of deutsche marks) for the years 1993 and 1994. The impressive rise of sales in the summer of 1994 was due to the fact that no division 1 prize (jackpot prize) was awarded for 10 consecutive weeks. This resulted in a superjackpot of more than 42 million DM (approximately $30 million U.S.).

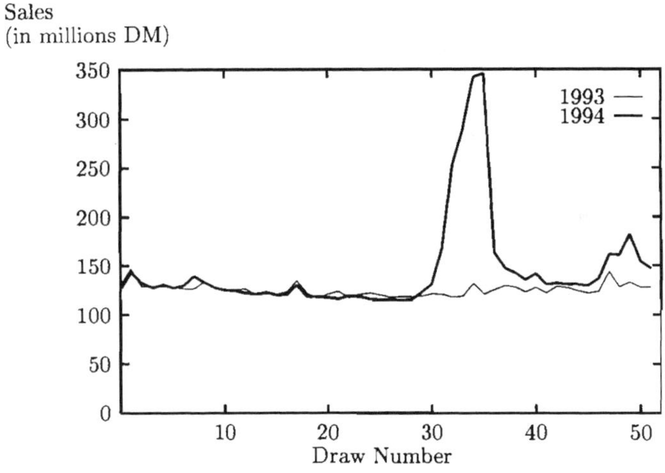

Figure 1.1 Weekly sales for the German Lotto 6/49: 1993 and 1994.

As state governments become more and more dependent on lottery revenues, there will be an increasing number of lotteries all over the world. There is little doubt that someday even a "Worldwide Ultimate Super Lotto" will be established. At the state level, interstate lottery players will look for the biggest jackpots and likely prefer state lotteries that award winners significantly more than the usual 50% of ticket sales or allow grand-prize winners the option of a lump sum payment. In sum, it seems quite likely that both lotteries and lotto players have a bright future ahead.

Chapter 2

The Rules

2.1 r/s Lotteries

According to Alexander Hamilton, the first Secretary of the Treasury of the United States, a successful lottery must have two basic characteristics: simplicity and low-priced tickets (see Sullivan, 1972, p. 7). Of course, simplicity is important because the scheme has to be easily understood by prospective participants. Low-priced tickets are also essential in order to encourage people to participate. In short, the goal is to present "fewer obstacles between hope and gratification" (see Sullivan, 1972, p. 10).

A common feature of all number lotteries is that you have to choose some numbers—hopefully, lucky ones. For example, the Massachusetts State Lottery offers three different number lotteries: *Megabucks*, *Mass Millions*, and *Mass Cash*. To play Megabucks, you have to select six numbers from 1 to 42. Playing Mass Millions involves choosing six numbers from 1 to 49. Finally, playing Mass Cash means picking five numbers from 1 to 35. The term *r/s lottery* can be used to characterize any number lottery in which you have to choose r numbers from among the numbers 1 to s. Thus, Megabucks is a 6/42 lottery, Mass Millions is a 6/49 lottery, and Mass Cash is a 5/35 lottery.

The following discussion addresses the common rules of a general r/s lottery. In a state that runs such a lottery, you buy a *play slip* or *betting slip* at a local retail outlet. On each play slip, there are several (usually between 5 and 10) *game boards* or *game panels.* Each game board contains boxes numbered from 1 to s; usually, s is between 30 and 54. Figure 2.1 shows several shapes of game boards from various states and national lotteries.

1	2	3	4	5	6	7
8	9	10	11	12	13	14
15	16	17	18	19	20	21
22	23	24	25	26	27	28
29	30	31	32	33	34	35
36	37	38	39	40	41	42
43	44	45	46	47	48	49

Germany 6/49

1	2	3	4	5	6
7	8	9	10	11	12
13	14	15	16	17	18
19	20	21	22	23	24
25	26	27	28	29	30
31	32	33	34	35	36
37	38	39	40	41	42
43	44	45			

Switzerland 6/45

1	2	3	4	5	6	7	8	9
10	11	12	13	14	15	16	17	18
19	20	21	22	23	24	25	26	27
28	29	30	31	32	33	34	35	36
37	38	39	40	41	42	43	44	45
46	47	48	49	50	51			

California 6/51

						1	2	3	3	5	6
7	8	9	10	11	12	13	14	15	16	17	18
19	20	21	22	23	24	25	26	27	28	29	30
31	32	33	34	35	36	37	38	39	40	41	42
43	44	45	46	47	48	49	50	51	52	53	54

New York 6/54

Figure 2.1 Typical shapes of game boards.

Using a heavy mark, you fill in the numbered boxes that correspond with your r (lucky numbers). (In some states, this mark must be vertical; in other states, it must be a cross; and in still others, the numbers must be circled.) In most states, r is five, six, or seven. Table 2.1 gives some examples of currently played r/s lotteries (as of July 1997).

Country/State	r	s	Name of lottery
Finland	7	39	Lotto Jokeri 7/39
Sweden	7	35	Sweden 7/35
Norway	7	35	Norway 7/35
New York	6	54	New York 6/54
Illinois	6	54	Illinois 6/54
California	6	51	California Super Lotto 6/51
Great Britain	6	49	British National Lottery 6/49
Florida	6	49	Florida 6/49
Maryland	6	49	Maryland 6/49
Massachusetts	6	49	Mass Millions 6/49
Germany	6	49	Lotto "6 aus 49"
Canada	6	49	Canadian Lotto 6/49
Spain	6	49	Lotería Primitiva
Pennsylvania	6	48	Pennsylvania Wild Card Lotto
Georgia	6	46	Lotto Georgia
Australia	6	45	Lotto 6/45 Australia
Switzerland	6	45	Swiss Lotto
Connecticut	6	44	Connecticut 6/44
Louisiana	6	44	Louisiana 6/44
Arizona	6	42	Arizona 6/42
Colorado	6	42	Colorado Lotto 6/42
Iowa	6	42	Iowa Super Cash Lotto
Kentucky	6	42	Kentucky 6/42
Massachusetts	6	42	Megabucks 6/42
New Zealand	6	40	New Zealand Lotto 6/40
District of Columbia	6	39	District of Columbia 6/39
Delaware	6	36	Delaware 6/36
Kansas	6	33	Kansas 6/33
California	5	39	California Fantasy 5
Connecticut	5	35	Connecticut 5/35
Massachusetts	5	35	Mass Cash 5/35
Florida	5	26	Florida New Fantasy 5

Table 2.1 Currently played r/s lotteries (July 1997).

Instead of doing a *manual pick*, in which you choose your lucky numbers yourself, you may do a *quick pick*. This means that a computer (with the help of a random number-generating algorithm) picks the numbers for you. In view of the fact that many lotto players bet on several of the same number patterns (see Chapter 5), and since lotto payoffs are often parimutuel (see Chapter 4), the strategy of playing a quick pick is a sensible way to ensure a bigger payoff if your numbers win (see Chapter 6).

Of course, there are rules (depending on the state or country) that apply if, for some reason, you choose fewer or more than r numbers on the game board. For example, if you choose fewer than six numbers in the New York Lotto 6/54 , the lottery will assign additional numbers at random, so as to process your application. If more than six numbers are circled in any game panel, however, just six will be selected at random by the lottery, and all the other numbers will be disregarded.

On the nights on which the winning numbers are drawn (once or sometimes twice a week), r numbered balls from s are chosen at random from a drawing machine to determine the winning combination. Usually, one additional ball is drawn as a so-called *bonus* or *supplemental number*. A notable exception is the Swedish Lotto 7/35, which has four bonus numbers. The draws are monitored by a lottery commission to ensure that each possible combination of numbers has the same chance of being selected. The drawing devices are designed to give the greatest possible degree of fairness. Before each draw, the numbered balls are examined very carefully and weighed to the nearest milligram. Any questionable ball is replaced, if doing so is advisable.

The advance of computer technology has significantly helped lottery operations around the world. Many lottery games are being converted to online systems that use the most advanced communication networks to link the central computer system of the lottery corporation with thousands of retailer terminals. Computerized lottery networks can handle several hundred thousand online plays per minute, each of which follows a definite path through the computer system, from the point of creation at the retailer's terminal to its final removal from the operating system files. Of course, the security of a fully computerized lottery system is paramount; extensive time and money have been invested in development to ensure the highest levels of performance and integrity.

2.2 Prize Structures

In awarding prizes, lotteries typically pay out decreasing amounts of money for specific number combinations that have decreasing odds (see Chapter 3). Usually, a ticket shares in the *jackpot* or *division 1 prize* if it correctly specifies all r winning numbers. An exception is the German Lotto 6/49, in which you have to match all the winning numbers, but, in addition, the last digit on the ticket identification number has to agree with a so-called *supernumber* chosen at random from the digits 0 through 9. This obstacle increases the odds for winning the jackpot by a factor of 10. A further exception is the multistate lottery Powerball 5/45 + 1/45, consisting of (as of July 1997) Arizona, Connecticut, Delaware, Idaho, Indiana, Iowa, Kansas, Kentucky, Louisiana, Minnesota, Missouri, Montana, Nebraska, New Hampshire, New Mexico, Oregon, Rhode Island, South Dakota, Washington, D.C., West Virginia, and Wisconsin. To win the Powerball 5/45 + 1/45 jackpot, you have to match all five winning numbers selected by a drawing machine from the numbers 1 through 45. In addition, you have to select one number from a separate 45-number field, and this number must match a so-called powerball drawn from a separate drawing machine.

To win a share of the *second* or *division 2 prize*, your ticket must correctly specify all but one of the winning numbers. In an r/s lottery, this means that you have to match any $r - 1$ of the r winning numbers. In cases in which a bonus number is drawn (for example, the Massachusetts Millions 6/49 Lotto, Canadian Lotto 6/49, and German Lotto 6/49), you get a share of the second prize only if you correctly specify any $r - 1$ of the r winning numbers plus the bonus number. (This is called the bonus prize in Canada.)

Additional consolation prizes are given for matching even fewer than $r-1$ of all r winning numbers. The prize-winning combinations vary from state to state and are even changed within states. Table 2.2 gives some examples of prize structures (as of July 1997). An entry k/r means that you have to match k of the r winning numbers, and if k is smaller than r, there is no further distinction between matching or not matching a possibly existing bonus number. A + sign indicates that you also have to match the bonus number. Having to match k of the r winning numbers without matching the bonus number is indicated by $k/r-$.

Lotto	Jackpot	2nd Prize	3rd Prize	4th Prize	5th Prize
Canada 6/49	6/6	5/6+	5/6–	4/6	3/6
Florida 6/49	6/6	5/6	4/6	3/6	–
California Super Lotto 6/51	6/6	5/6	4/6	3/6	–
New York 6/54	6/6	5/6	4/6	3/6+	–
Ohio 5/37	5/5	4/5	3/5	2/5	–
Massachusetts Mass Cash 5/35	5/5	4/5	3/5	–	–

Table 2.2 Prize structures of r/s lotteries (July 1997).

And finally, to the all-important question, **How much do I win when I win**? The answer is, **It all depends**.

This point will be discussed in detail in Chapter 4. However, a general rule that applies to many r/s lotteries and most prize levels, the prize pool available for a given prize level (which varies weekly) is split among all winners of that prize level. This suggests that the only strategy for getting a better long-term return on investment when playing lotto is to play combinations that are unpopular with other players (see Chapters 4 to 7).

Another rule that applies to most U.S. r/s lotteries is that larger prizes are paid in installments. For example, the California lottery pays grand prizes of $1 million or more in equal installments over 20 years. These prizes are funded by U.S. Treasury zero coupon bonds, which are purchased for each winner with annual maturity dates that result in the full prize amount. For each grand prize, the first payment is made a few weeks after the prize is claimed, and a bond is bought for each subsequent annual payment.[1] However, in other states, such as Colorado, jackpot winners have their choice of either cash or annuity prizes.

Since playing lotto, like many other things, means submitting to a commission's rules and regulations, you should contact your state's lottery commission to answer any specific questions concerning your r/s lottery.

[1] California State Lottery, *Fact Sheet* (Revision 1–5–95).

Chapter 3

The Odds

3.1 Chances, Odds, and Probabilities

What are your chances of winning prizes in a general r/s lottery? What do these chances really mean?

The mathematical term for *chance* is *probability*. For example, when you cast a fair die, the probability that a specific face will show up is 1/6, which means 1 out of 6 equally likely outcomes. In other words, the chance is 1 in 6 or the odds are 6 to 1. Thus, higher odds correspond to smaller probabilities (chances) and vice versa.

More generally, the probability of some chance event, denoted by A, is p (a value between 0 and 1). For instance, suppose you have a long series of chance experiments that are carried out under the same conditions and independently of each other. Suppose further that in each of these trials, the event A may or may not occur. Then, in the long run, the number of occurrences of A, divided by the number of trials performed (that is, the *relative* frequency of occurrences of A), will tend to match the value p. This means that if the number of experiments grows indefinitely, the event A will occur in about $p \times 100\%$ of all cases. Put differently, the chance of A is 1 in $1/p$ or the odds of A are $1/p$ to 1.

To look at odds and probabilities from another perspective, note that in the long run, the odds value $1/p$ is given approximately by the total number of experiments divided by the number of all those experiments in which event A occurs. That is, $1/p$ may be regarded as the average number of trials needed in the long run to produce one occurrence of event A. The mathematical term corresponding to this average number is *expected value*. Thus, if the probability of event A is p, the expected value of the random variable that represents the number of experiments needed to observe the first occurrence of A is given by the reciprocal (odds) value $1/p$.

The chances of winning a specific level of prize in a general r/s lottery depend on three factors:

1. r and s

2. the prize structure

3. the number of tickets (combinations) you play

The number of possibilities of choosing r out of s numbers is given by the quotient

$$\frac{s \times (s-1) \times (s-2) \times \cdots \times (s-r+2) \times (s-r+1)}{1 \times 2 \times 3 \times \cdots \times (r-1) \times r}$$

(see Chapter 8, Section 8.1). In particular, there are

$$\frac{49 \times 48 \times 47 \times 46 \times 45 \times 44}{1 \times 2 \times 3 \times 4 \times 5 \times 6} = 13,983,816$$

possible winning combinations in a 6/49 lottery and

$$\frac{54 \times 53 \times 52 \times 51 \times 50 \times 49}{1 \times 2 \times 3 \times 4 \times 5 \times 6} = 25,827,165$$

possible winning combinations in a 6/54 lottery.

So, if you play one ticket in a 6/49 lottery, the probability of winning the next draw's jackpot is $1/13,983,816$; that is, the jackpot odds are 13,983,816 to 1. Of course, the jackpot odds will decrease if you play more tickets. For example, playing 10 different tickets gives you a chance of 10 in 13,983,816 of winning the jackpot. In other words, playing 10 different tickets decreases the jackpot odds by the factor 1/10. (Formulae for calculating the probabilities of winning prizes of specified levels in a general r/s lottery are given in Chapter 8, Section 8.1.)

In the discussion that follows, the *odds values* (reciprocals of probabilities) will be given. For the various prize levels of the California Super Lotto 6/51, these values are shown in Table 3.1.

Prize Level	Odds of Winning
6/6	18,009,460 to 1
5/6	66,702 to 1
4/6	1,213 to 1
3/6	63.5 to 1

Table 3.1 Odds of winning the various prize levels for the California Super Lotto 6/51.

	k					
s	6	5+	5−	5	4	3
54	25,827,165	4,304,528	91,586	89,678	1,526	74.7
51	18,009,460	3,001,577	68,218	66,702	1,213	63.5
49	13,983,816	2,330,636	55,491	54,201	1,032	56.7
48	12,271,512	2,045,252	49,884	48,696	950	53.4
46	9,366,819	1,561,137	40,029	39,028	801	47.4
45	8,145,060	1,357,510	35,724	34,808	733	44.6
44	7,059,052	1,176,509	31,798	30,961	669	41.8
42	5,245,786	874,298	24,980	24,286	555	36.7
40	3,838,380	639,730	19,386	18,816	456	32.1
39	3,262,623	543,771	16,993	16,478	412	29.9
36	1,947,792	324,632	11,194	10,821	299	24.0
33	1,107,568	184,595	7,100	6,837	210	18.9

Table 3.2 Odds values for matching k winning numbers in a $6/s$ lottery.
Note: + means "with bonus number," and – means "without."

Table 3.2 gives the odds values for matching k numbers in a general "choose 6 out of s lottery" for various values of s. A + sign means "with bonus number," and a − sign means "without a bonus number." (For brevity, the appendage "to 1" has been omitted.)

To get a real feeling for these odds, consider the odds of events from random experiments that you can conduct yourself. For example, when you cast three fair dice simultaneously, the odds are 216 (= 6 × 6 × 6) to 1 that all of them will show 6s. More generally, Table 3.3 illustrates for several values of n the odds of getting all 6s when casting n fair dice simultaneously. (Again, the appendage "to 1" has been omitted.)

n	Odds	n	Odds
2	36	6	46,656
3	216	7	279,936
4	1,296	8	1,679,616
5	7,776	9	10,077,696

Table 3.3 Odds that all of n dice will show 6s when thrown simultaneously.

Likewise, Table 3.4 shows for several values of n the odds that all of n fair coins will show "heads" when tossed simultaneously.

n	Odds	n	Odds	n	Odds
5	32	12	4,096	19	524,288
6	64	13	8,192	20	1,048,576
7	128	14	16,384	21	2,097,152
8	256	15	32,768	22	4,194,304
9	512	16	65,536	23	8,388,608
10	1,024	17	131,072	24	16,777,216
11	2,048	18	262,144	25	33,554,432

Table 3.4 Odds that all of n fair coins will show "heads" when tossed simultaneously.

3.2 How Many Years Will It Take to Win?

Another way to put the odds of winning lotto into perspective is to realize how many years it will take, on average, to win a prize of a specified category. Of course, this will depend on the number of combinations that you play. Table 3.5 shows how many years it will take, on average, to match k winning numbers (with or without a bonus number) for a $6/s$ lottery, if you play one ticket per draw and the draws are conducted weekly.

	k							
s	6	5+	5−	5	4	3+	3−	3
54	494,300	82,383	1,753	1,716	29.2	22.9	1.5	1.4
51	344,679	57,446	1,306	1,277	23.2	18.2	1.3	1.2
49	267,633	44,606	1,062	1,037	19.8	15.5	1.2	1.1
48	234,862	39,144	955	932	18.2	14.3	1.1	1.0
46	179,269	29,878	766	747	15.3	12.1	1.0	0.9
45	155,886	25,981	684	666	14.0	11.1	0.9	0.9
44	135,102	22,517	609	593	12.8	10.1	0.9	0.8
42	100,398	16,733	478	465	10.6	8.4	0.8	0.7
40	73,462	12,244	371	360	8.7	7.0	0.7	0.6
39	62,443	10,407	325	315	7.9	6.3	0.6	0.6
36	37,278	6,213	214	207	5.7	4.6	0.5	0.5
33	21,198	3,533	136	131	4.0	3.3	0.4	0.4

Table 3.5 Expected waiting times (years) for matching k winning numbers in a $6/s$ lottery.
Note: + means "with bonus number," and − means "without."

Clearly, any given expected waiting times will be shorter if you choose more than one combination and/or if there is more than one draw per week. If you play m different games, of which no two share more than two numbers, and if there are w draws per week, each entry in Table 3.5 will have to be divided by $m \times w$. As an example, suppose that you play six different games in the New York Lotto 6/54, which has two draws per week. The expected waiting time to win a fourth prize (3+) is given by $22.9/(6 \times 2) \approx 1.9$; thus, you will have to wait nearly two years, on average.

You might conclude that these average waiting times of many thousands of years don't have much significance for a single lotto player, who will not be able to play for more than 50 or 60 years. Any player will be much more interested in his or her chance of winning the jackpot within a reasonable period of time—hopefully, early enough in life to enjoy it properly! Table 3.6 provides that information, showing for the values $j = 10, 20, \ldots, 80$ and several values of s the odds of winning the jackpot in a $6/s$ lottery at least once within a period of j years, assuming that the draws are conducted weekly and you play 10 different combinations per week.

	j							
s	10	20	30	40	50	60	70	80
54	4,939	2,472	1,648	1,236	1,028	824	707	618
51	3,444	1,724	1,149	863	717	575	493	431
49	2,674	1,339	892	670	557	447	383	335
48	2,347	1,175	783	588	489	392	336	294
46	1,791	897	598	449	373	299	257	225
45	1,558	780	520	390	325	260	223	195
44	1,350	676	451	338	281	226	193	169
42	1,004	502	335	251	209	168	144	126
40	734	368	245	184	153	123	105	92
39	624	313	209	157	130	105	90	79
36	373	187	125	94	78	63	54	47

Table 3.6 Odds of winning the jackpot in a $6/s$ lottery at least once within j years (if draws are conducted weekly and 10 different combinations are played per week).

For example, suppose that you are playing 10 combinations in the California Super Lotto 6/51, which has two draws per week. The odds of winning the jackpot at least once within 20 years are 863 to 1. (These correspond to 40 years if the draws are held weekly.) Note that these odds are nearly the same as those having all three dice show 6s or both of two coins show "heads" when cast simultaneously. (These odds are given by 1 in 864 [$= 6 \times 6 \times 6 \times 2 \times 2$].)

It is clear that these odds get worse if you play less than 10 combinations. If you play 5, for example, your odds of winning the jackpot within j years will be about double the corresponding values in Table 3.6 (that is, multiply by 2). If you play 2 combinations, you have to multiply each entry in Table 3.6 by 5, and if you play just 1 combination, by 10.

3.3 You Can't Beat the Odds!

You might become disillusioned if you do these calculations for your state's specific lotto. Moreover, even with this information, you cannot beat the odds with any system (that is, by charting numbers of past drawings), because lottery numbers are randomly selected. This point is of paramount importance, especially today, when computer programs facilitate the discovery of various interesting number patterns, such as the following:

- *frequency-hit patterns*, which display how often a number came up in the history of the draws
- *last-time-hit patterns*, which show when a number was last selected
- *skip-and-hit patterns*, which provide information about a number's absence and appearance sequence during the history of the lottery
- *repeating-pairs patterns*, which display the number of occurrences of a pair of numbers as winning numbers
- *repeating-triples patterns*, which provide information on the number of occurrences of a triple of numbers as winning numbers

Bear in mind that the drawing device doesn't care about past winning combinations; what happened in the past has no relevance for future draws. Thus, analyzing the results of past lotteries to look for interesting patterns of numbers will do you no good. Even so, many lotto players continue to look for number patterns, certain that they will find something if they just search long and hard enough.

Trying to find patterns in past winning combinations is not only absolutely useless but also extremely dangerous, if such patterns are used to select combinations for play. (Consider that many other players might do the same thing.) Chapter 5 will show why playing strategies based on frequency-hit patterns or last-time-hit patterns are extremely foolish styles of play. In fact, the *only* benefit from previous drawings is to know the past winning combinations in order to avoid them because of their widespread popularity among other players (see Chapter 5, Section 5.2).

Only one strategy will help you win more *when you are lucky to have picked a winning combination*: Avoid popular combinations so you won't have to share jackpots with many other players *when you win.*

Chapter 4

Learning from Prize Payouts

4.1 The Parimutuel Payout System

Now that you know about the rather unfavorable odds of winning at lotto, you may wonder where all the money invested in lottery tickets goes. A common feature of all lotteries is that only a certain percentage—usually, between 45% and 55%—of the draw (ticket) sales is allocated to the prize fund. The rest of the draw sales is divided among a number of supposedly good causes: to the government in lottery duties ("painless taxes"), to the retailers in commissions, and toward operating costs for the lottery corporations (see Chapter 1). This information may or may not be important to individual players. How the prizes are calculated, however, should be very important to all players.

Nearly all lotteries follow either a purely *parimutuel payout system* or a parimutuel system *combined with a fixed payout* for the lowest prize category. If the payout for a prize category is determined on a parimutuel basis, the total prize money available for that category will be shared between the winners in that category.

The British National Lottery 6/49, for example, allocates 45% of the draw sales of each game to a prize fund for that draw (as of July 1997). Each match 3 prize winner receives a fixed prize of £10 (British pounds), except in the extremely unlikely event that the prize fund is too small to give each winner £10. Should this occur, the entire prize fund for that draw would be divided equally between the total number of winners in all categories.

The prize money allocated to each of the higher prize categories is shared between the winners in that category. Following payment of the match 3 prizes, the remaining prize fund, now termed the *pools fund*, is divided as follows:

- match 4 numbers = 22%
- match 5 numbers = 10%
- match 5 numbers plus the bonus number = 16%
- match all 6 numbers = 52%

If no one has matched all six numbers, the jackpot prize is rolled over to the next draw. A specific feature of the British National Lottery is that the number of jackpot rollovers is restricted. If no one wins the jackpot after four consecutive draws, the final jackpot fund will be shared among the players who have matched five of the six winning numbers plus the bonus number in that draw.

Match	Prize
6	£14,560,000
5+	£4,480,000
5	£2,800,000
4	£6,160,000

Table 4.1 Sample prize distribution for the British National Lottery.

To illustrate this payout system in round numbers, assume that the total stake is £100,000,000 for some draw of the British National Lottery. This means that the total prize fund is £45,000,000 (45% of the total stake). If there are 1,700,000 match 3 prize winners, £17,000,000 is taken from the prize fund, leaving a pools fund of £28,000,000. According to the percentages just given, the pools fund is divided as shown in Table 4.1. According to the parimutuel payout system, each particular prize fund will be divided among the winners of that prize category. Thus, if 100 players have matched all six numbers, each will have to be content with a fairly modest jackpot of £145,600. This actually happened on January 14, 1995, when 133 players hit the winning combination 7, 17, 23, 32, 38, 42, giving each of them a prize of £122,510. Only four weeks later, a single player hit the winning numbers 1, 7, 37, 38, 42, 46 and was happy to receive the whole jackpot of £10,350,387.

For some reason (to be discussed later), the combination 7, 17, 23, 32, 38, 42 was much more attractive to British lotto players than the numbers 1, 7, 37, 38, 42, 46. This points to an important guideline for increasing a lottery win: Know the typical player's behavior in order to learn and avoid popular number combinations. The main problem is to assess the *degree of popularity* of a given number combination. Again, popular combinations mean shared jackpots and thus lower individual winnings.

4.2 Examples of Extreme Payouts

Figure 4.1 illustrates the possible fluctuation of prizes. The left-hand game ticket shows the combination that gave the largest match 5 and at the same time the largest match 4 payout in the British National Lottery in 1996. This combination is marked with a ☺ in order to indicate its low popularity and thus high payoff quality. The right-hand ticket in Figure 4.1 deserves a ☹ because its high popularity (low payoff quality) resulted in a minimum match 5 (minimum match 4) payout of only £454 (£33) in 1996.

Likewise, each of the three combinations displayed in Figure 4.2 gave the same low minimum match 4 prize amount of £33 in 1996.

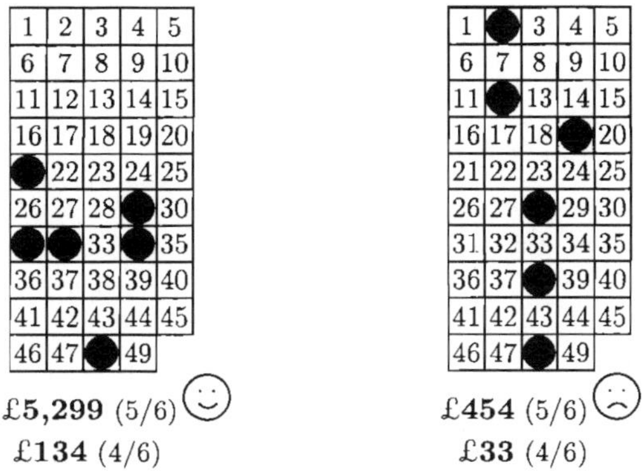

Figure 4.1 Extreme match 5 and match 4 payouts in the British National Lottery (1996).

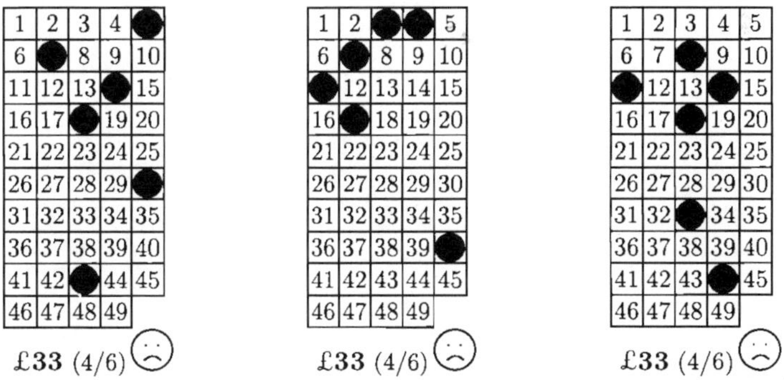

Figure 4.2 Minimum match 4 payouts in the British National Lottery (1996).

Figures 4.1 and 4.2 indicate several popular types of player behavior:

• *Playing on visual impulse* means that players often try to scatter numbers in a way that *seems random* to them. However, the resulting combinations are seldom truly random. There is a widespread misconception that randomness means that numbers are spread out evenly over the game board. On the contrary, truly random combinations tend to have clusters of neighboring numbers and also large gaps between numbers (see Chapter 8, Section 8.7).

• *Choosing nice-looking geometric patterns* is one of the most foolish styles of play (see Chapter 5, Section 5.6).

• *Playing on a calendar-oriented basis* means that players select birthdays or other (typically low) lucky numbers, such as 3, 7, and 12.

Thus, the left-hand combination in Figure 4.1 is unpopular because the numbers are neither spread out evenly over the game board nor selected on a calendar-oriented basis. Compared with the right-hand combination in Figure 4.1, the left-hand combination (21, 29, 31, 32, 34, 48) simply looks ugly. In comparison, the three combinations in Figure 4.2 are visually appealing and/or emphasize low numbers.

The same kind of player behavior is illustrated in Figure 4.3.

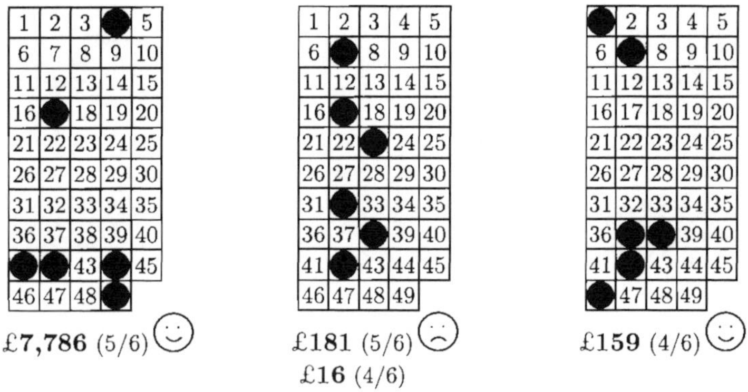

Figure 4.3 Extreme match 5 and match 4 payouts in the British National Lottery (1995).

The two outer tickets—best of the year 1995 match 5 and match 4—look ugly. There are large gaps between numbers, clusters of neighboring numbers, and only a few low numbers. On the other hand, the nearly regular-looking center ticket was the worst-paying combination of 1995. Apart from giving extreme low match 5 and match 4 payoffs, it produced a record number of 133 jackpot winners (see p. 24).

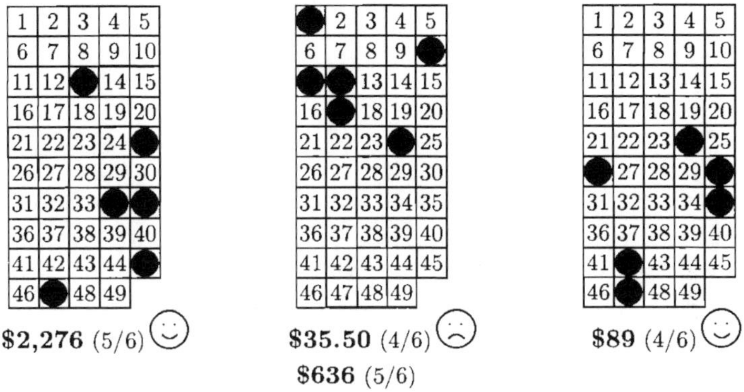

Figure 4.4 Extreme match 5 and match 4 payouts in the Florida State Lottery (1996).

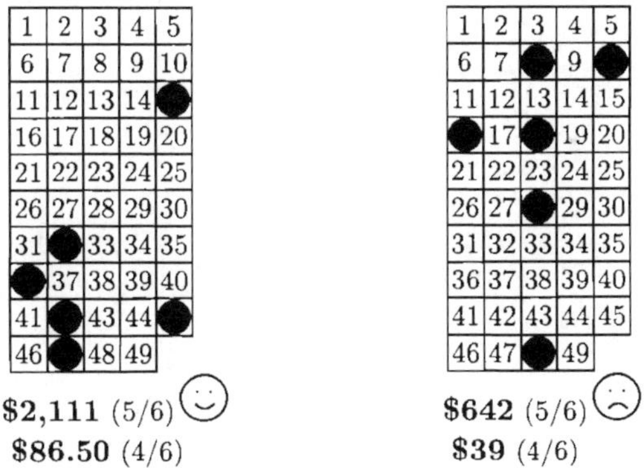

Figure 4.5 Extreme match 5 and match 4 payouts in the Florida State Lottery (1995).

For further illustration, Figures 4.4 and 4.5 show (respectively) the combinations with the largest and smallest match 5 and match 4 prize amounts for the Florida State Lottery 6/49 in 1996 and 1995. It is striking to see that lotto players in Florida behave much the same way as British lotto players. Thus, avoiding low numbers (the outer tickets in Figure 4.4), perhaps in combination with a large gap (the left-hand ticket in Figure 4.5), is likely to yield a high payout, whereas playing only low numbers (the middle ticket in Figure 4.4) or a nearly symmetric ticket (the right-hand ticket in Figure 4.5; note that replacing the number 10 with 20 gives the visual impression of a cross) is bound to give a very bad return on investment.

Additional examples are provided by Figures 4.6 and 4.7, which (respectively) show the tickets with the largest and smallest match 5 and match 4 prize amounts for the German Lotto 6/49 in 1996 and 1995.

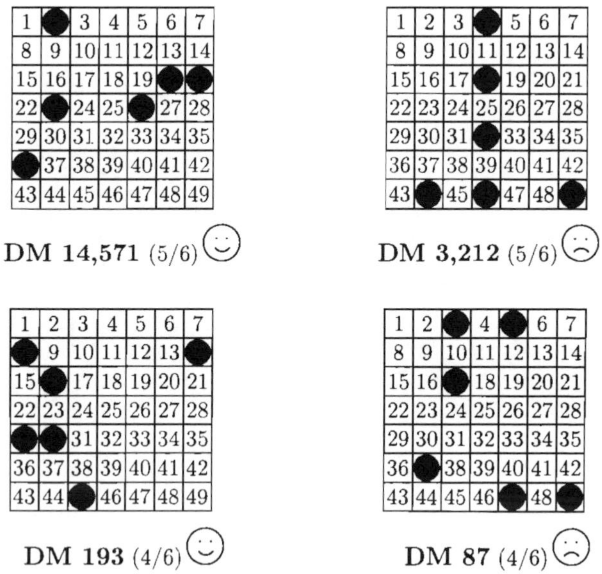

Figure 4.6 Extreme match 5 and match 4 payouts in the German Lotto 6/49 (1996).

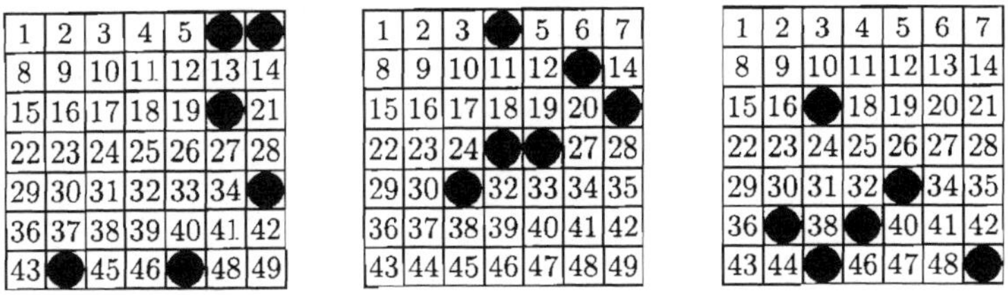

DM 12,735 (5/6) **DM 1,773** (5/6) **DM 83** (4/6)
DM 193 (4/6)

Figure 4.7 Extreme match 5 and match 4 payouts in the German Lotto 6/49 (1995).

Consistent with lotto players from Great Britain and Florida, German lotto players like visually attracting combinations, such as the right-hand tickets in Figures 4.6 and 4.7. Note that the upper-right-hand ticket in Figure 4.6 is nearly a perfect reversed T, which would be obtained by substituting 48 for 49. Likewise, the right-hand ticket in Figure 4.7 is nearly a perfect zigzag (replace 17 by 41).

These low-payoff tickets provide further evidence that you should stay away from nice-looking combinations. Indeed, playing geometric patterns or trying to approximate ideal geometric patterns (such as lines, diagonals, crosses, zigzags, etc.) is one of the most foolish styles of play (see Chapter 5, Section 5.6). Whereas the left-hand tickets in Figures 4.6 and 4.7 brought high payoffs because of their ugliness, the middle ticket in Figure 4.7 deserves special attention. Consider the extremely low match 5 payout of only DM 1,773, particularly in comparison with the second-lowest payoff of DM 3,597 and typical match 5 payoffs between DM 5,000 and 11,000. This comparison reveals yet another widespread foolish behavior among lotto players: namely, choosing winning combinations from past draws (see Chapter 5, Section 5.2). In the combination 4, 13, 21, 25, 26, 31, five of the six numbers were drawn as winning numbers only seven weeks before (see p. 44)!

As a further example of popular and unpopular styles of play, Figure 4.8 displays the extreme match 5 and match 4 payoffs for the Canadian Lotto 6/49 in 1994.

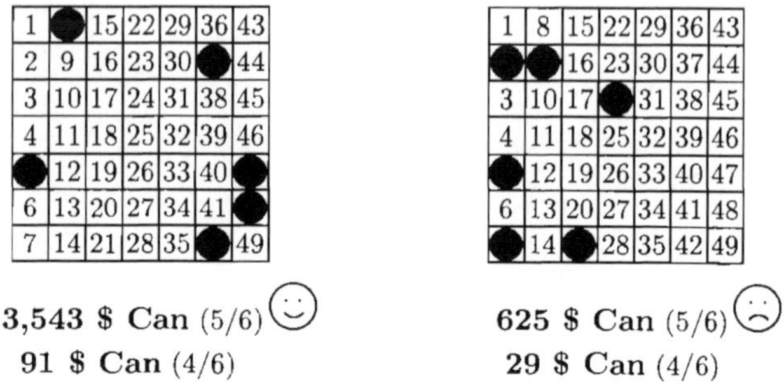

Figure 4.8 Extreme match 5 and match 4 payoffs in the Canadian Lotto 6/49 (1994).

Playing on a calendar-oriented basis in this lotto means choosing numbers mainly from the left-hand part of the game board, since in some Canadian provinces, the 49 numbers are set columnwise from top to bottom. This kind of player behavior is reflected by the right-hand combination in Figure 4.8. The left-hand ticket in Figure 4.8, however, has all the features of a high-payoff combination, including a large gap (between 8 and 31), only two numbers in the "birthday range" from 1 to 31, and a cluster of three numbers (42, 47, 48).

As a final case in point, consider the extreme match 5 and match 4 prize amounts for the California Super Lotto 6/51 for the year 1994, as shown in Figure 4.9. The maximum prize payout for the division 2 prize (5 of 6) was $2,796, whereas the minimum prize payout for that division was only $621. Likewise, the range of prize payouts for the division 3 category (4 of 6) was from a minimum of $60 to a maximum of $108. It is striking to see that, in complete accordance with the previous findings from lotteries in Great Britain, Florida, Canada, and Germany, the low-payoff tickets are characterized by low numbers (and, in particular, the lucky numbers 7 and 12), whereas the high-payoff tickets contain few low numbers.

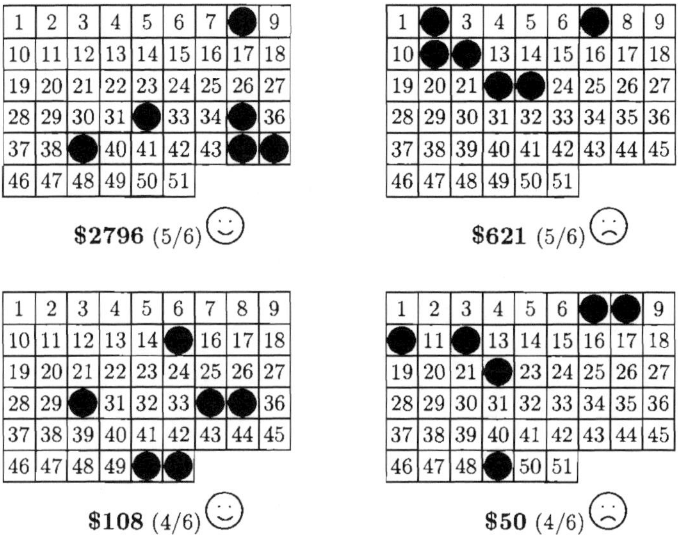

Figure 4.9 Extreme match 5 and match 4 payouts for the California Super Lotto 6/51 (1994).

4.3 Seeing Through Scatterplots

You may ask whether there is some kind of connection between avoiding low numbers and receiving high payoffs. To answer this question, examine Figures 4.10 and 4.11.

Figure 4.10 shows a scatterplot of 105 points, each point representing one draw in the California Super Lotto 6/51 in 1994. For each draw, the sum of the winning numbers is plotted against the match 4 prize for that draw. Obviously, there is some sort of relationship between the sum of the winning numbers, X, and the match 4 payout, Y. At least on the average, larger values of X correspond to larger values of Y; thus, it is tempting to try to predict Y from X by using a suitable straight-line relationship. Indeed, there is a fairly high correlation (see Chapter 8, Section 8.10) between the sum of the drawn numbers and the match 4 prize payout.

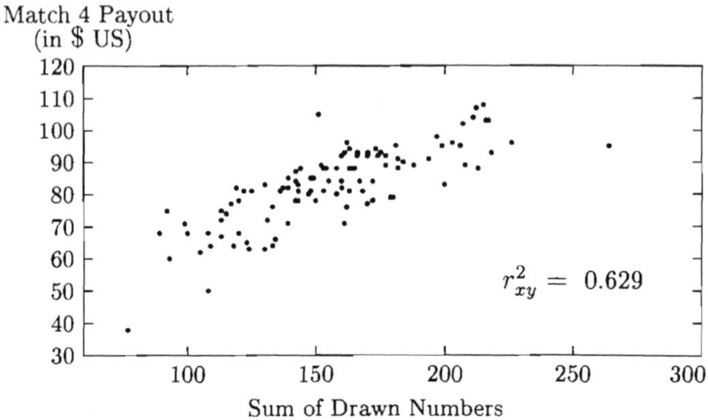

Figure 4.10 Scatterplot sum of drawn numbers vs. match 4 payout.

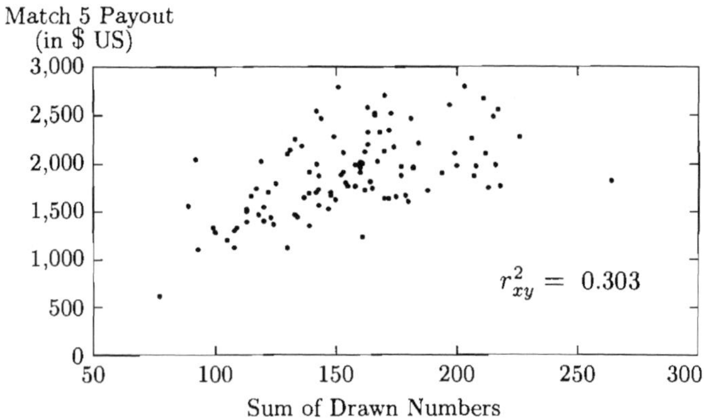

Figure 4.11 Scatterplot sum of drawn numbers vs. match 5 payout.

Figure 4.10 illustrates two widespread styles of play: namely, playing on a calendar-oriented basis and preferring special small lucky numbers like 3, 7, and 12. Although there is a statistical relationship between the sum of the drawn numbers and the match 4 prize payout, you would draw a completely wrong conclusion by choosing the combination with the largest sum of numbers possible (the numbers 46, 47, 48, 49, 50, 51). Betting on this combination is one of the most foolish styles of play (see Chapter 5, Section 5.1).

Figure 4.11 also shows a scatterplot of the sum of the drawn numbers against the match 5 payout. Compared with that of Figure 4.10, the correlation in 4.11 between the sum of the winning numbers and the match 5 prize amount is less pronounced but still noticeable.

In many of the examples considered so far, the extreme match 5 and match 4 payouts, taken over a time period of one year, were attained for the same combination of numbers (see Figures 4.1, 4.5, and 4.8). Since payouts depend on how popular combinations of winning numbers are among players, a high match 4 payout

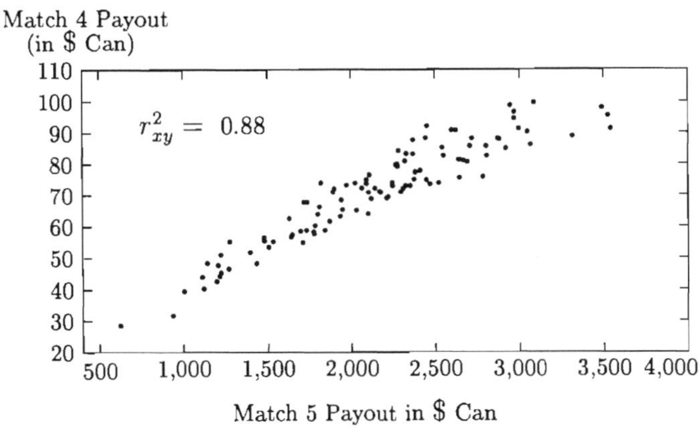

Figure 4.12 Scatterplot match 5 vs. match 4 payout (in $ Can).

can be expected to go along with a high match 5 prize and even a large share of the jackpot. Conversely, a low match 4 payout can be expected to go along with a low match 5 prize and even a small share of the jackpot.

To illustrate both sides of this phenomenon, Figure 4.12 gives a scatterplot of the match 5 payout and match 4 payout for the Canadian Lotto 6/49 in 1994.

The strong relationship between the match 5 and match 4 payouts, as evidenced by a very high correlation, becomes even more pronounced when we consider the *natural logarithm*, or ln(.), of the payouts (see Chapter 8, Section 8.14). The resulting scatterplot is shown in Figure 4.13.

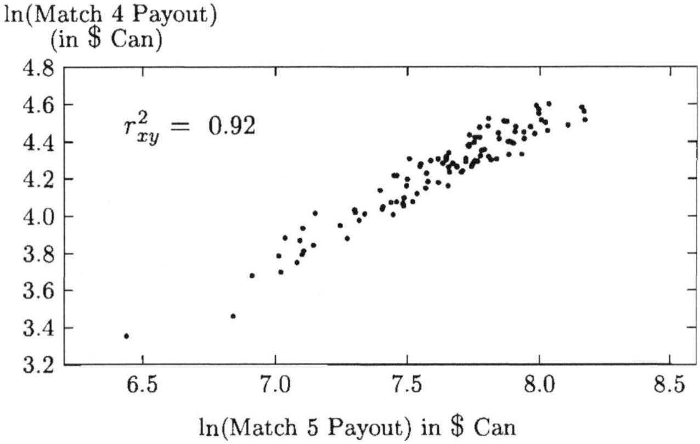

Figure 4.13 Scatterplot ln(match 5) vs. ln(match 4) payout.

At this point in the chapter you may wonder why, with one notable exception on p. 24, the chances of winning jackpot prize payouts have not been discussed. The reason is that there are so few jackpot winners; consequently, calculating statistics for such winners is not justified. Nonetheless, you will have your best chance of winning a whole jackpot by avoiding the many foolish styles of play discussed (see Chapter 5). Also keep in mind that a high match 5 prize amount is likely to go along with a high payout for the next higher prize division, which for the Canadian Lotto 6/49 means matching five of the winning numbers plus the bonus number. This is illustrated in Figure 4.14, which gives a scatterplot of the natural logarithms of the match 5 and match 5+ payouts for 1994.

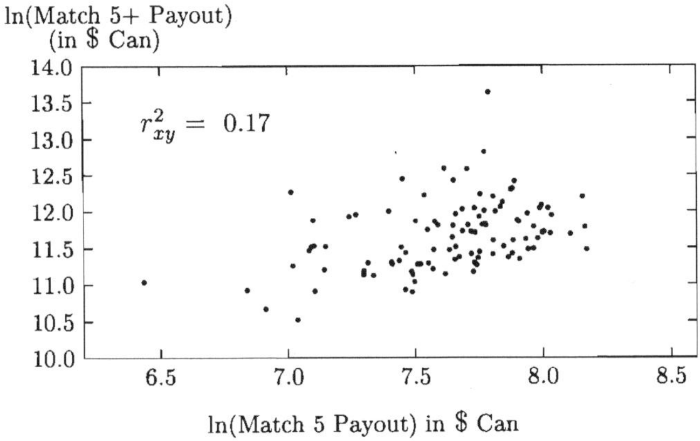

Figure 4.14 Scatterplot of the ln(match 5) vs. ln(match 5+) payout.

Having read this chapter, you are aware of the large range of prize payouts that is possible. And upon viewing the many low-payoff tickets presented, you may perhaps be questioning some of your own favorite number combinations. The intelligent player does what he or she can to ensure receiving a high prize payout on those rare winning occasions. The foolish player may as well donate his or her money to the state.

Chapter 5

Foolish Styles of Play

What is the lotto player's nightmare? Having to split a jackpot with hundreds or even thousands of other players on a presumably once-in-a-lifetime winning occasion. Reading this chapter can help you avoid such a nightmare. The most important thing you should learn is how *not* to play lotto. Remember that at least for higher-division prizes, lotto payoffs are made on a parimutuel basis; therefore, avoiding other players' choices of numbers will increase your own chances of winning big. But what should you do if you don't know what to avoid? Here is the answer, which can be called, for lack of a better term, the *categorical imperative for lotto players*:

Never choose your numbers according to a rule that might be followed by other players.

This chapter is full of examples that will open your eyes as to what may happen if you act against this principle. Indeed, acting against the categorical imperative for lotto players is the most foolish style of play.

You may wonder what is meant by the word *rule*. A *rule* is any simple prescription for choosing lotto numbers that, when applied, leads to only one or relatively few number combinations. To give you an example of a rule, recall that a *square number* is one that can be written in the form $k \times k$ for some number k. The simple rule "Play the first six square numbers" leads to only one combination: namely, 1, 4, 9, 16, 25, 36.

5.1 Choosing Arithmetic Progressions

Another rule for choosing numbers would be "Start with some number, and always add 3 to get the next number." In a 6/49 lottery, this rule leads to 1 of these 34 combinations:

1,	4,	7,	10,	13,	16
2,	5,	8,	11,	14,	17
3,	6,	9,	12,	15,	18
...	...	...	...	...	...
33,	36,	39,	42,	45,	48
34,	37,	40,	43,	46,	49

More generally, a combination formed in this way—that is, by choosing a smallest value a and always adding the same value d to obtain the next number—is called an *arithmetic progression*. It is not difficult to understand that only 216 different arithmetic progressions are possible in a 6/49 lottery (44 having $d = 1$, 39 having $d = 2$, 34 having $d = 3$, and so on). Thus, "Choose some arithmetic progression" is also a rule. In fact, as you will see, it is one of the worst.

To give you an impression of the most foolish styles of play, look at Tables 5.1 through 5.4. Table 5.1 lists the frequencies of the 10 most popular number combinations chosen for the October 29, 1988, draw of the California Lotto 6/49 (see Kadell & Ylvisaker, 1991). In the same way, Table 5.2 shows the frequencies of the 10 most popular combinations from a total of 16,862,596 chosen for the February 10, 1990, draw of the Swiss Lotto 6/45 (see Riedwyl, 1990, p. 46).

Combination						Frequency
7	14	21	28	35	42	16,771
1	2	3	4	5	6	15,415
1	4	15	22	26	46	11,821
5	10	15	20	25	30	10,789
3	8	13	18	23	28	9,034
8	16	24	32	40	48	6,823
3	13	23	33	43	48	6,629
5	10	13	22	27	33	6,340
1	6	11	16	21	26	6,093
18	20	29	32	33	35	5,456

Table 5.1 Frequencies of the 10 most popular number combinations for the October 29, 1988, draw of the California Lotto 6/49.

Combination						Frequency
6	11	16	21	26	31	24,120
1	8	15	22	29	36	24,009
7	14	21	28	35	42	17,913
8	14	17	25	31	36	12,008
3	9	15	21	27	33	11,297
1	2	3	4	5	6	10,637
12	17	22	27	32	37	10,170
6	12	18	24	30	36	8,663
18	23	28	33	38	43	8,354
4	10	16	22	28	34	8,147

Table 5.2 Frequencies of the 10 most popular number combinations for the February 10, 1990, draw of the Swiss Lotto 6/45.

Table 5.3 shows the frequencies of the 10 most popular number combinations from a total of 6,803,090 chosen for the October 16, 1993, draw of the Federal State Baden–Württemberg Lotto 6/49 of Germany (see Bosch, 1994, p. 215).

Combination						Frequency
7	13	19	25	31	37	4,004
7	14	21	28	35	42	3,817
5	27	34	35	37	49	3,698
1	2	3	4	5	6	3,249
4	11	18	25	32	39	2,821
13	19	25	31	37	43	2,335
6	12	18	24	30	36	2,288
9	17	25	33	41	49	2,227
1	9	17	25	33	41	2,116
8	16	24	32	40	48	2,097

Table 5.3 Frequencies of the 10 most popular number combinations for the October 16, 1993, draw of the Federal State Baden–Württemberg Lotto 6/49.

And for a final example, Table 5.4 gives the frequencies of the 10 most popular of a total 12,707,845 combinations chosen for the August 27, 1994, draw of the Finnish Lotto 7/39.

Combination							Frequency
1	2	3	4	5	6	7	5,066
5	10	15	20	25	30	35	3,225
1	7	13	19	25	31	37	3,123
3	9	15	21	27	33	39	2,770
2	8	14	20	26	32	38	2,030
11	15	18	30	34	35	36	1,492
6	11	16	21	26	31	37	1,421
1	7	10	16	24	29	32	1,042
5	8	10	11	21	29	36	884
8	12	13	23	24	28	33	868

Table 5.4 Frequencies of the 10 most popular number combinations for the August 27, 1994, draw of the Finnish Lotto 7/39.

Now take a closer look at Tables 5.1 through 5.4. It should be apparent that, in each case, the most popular number combination is an arithmetic progression. The most popular combinations from the four tables are shown in Figure 5.1. Furthermore, 6 of the California top 10 combinations are arithmetic progressions, and the rank 7 combination (3, 13, 23, 33, 43, 48) may be regarded as a *broken arithmetic progression.* (In a 6/53 lottery, 48 would presumably have been replaced by 53.) For the German and Swiss lotteries, all but 1 of the top 10 tickets are arithmetic progressions. In sum, each of the most popular tickets was chosen at a frequency thousands of times above average.

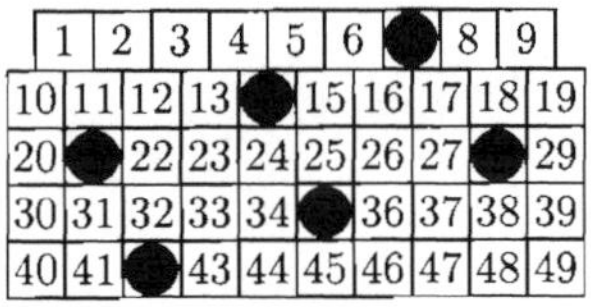

California Lotto 6/49
October 29, 1988

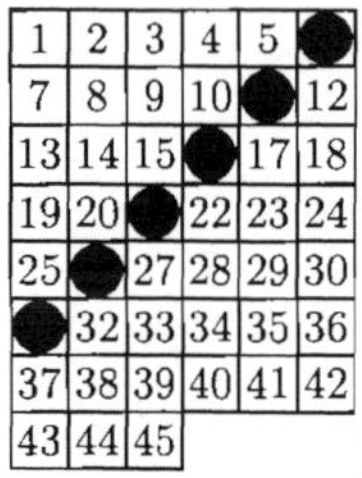

Swiss Lotto 6/45
February 10, 1990

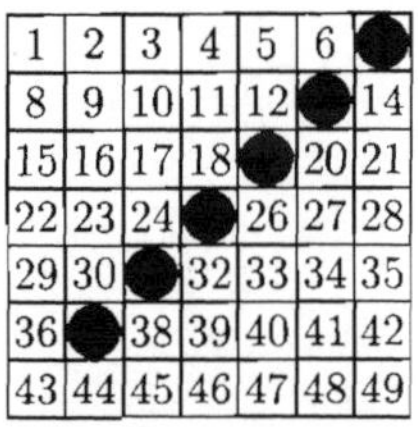

Baden–Württemberg Lotto 6/49
of Germany
October 16, 1993

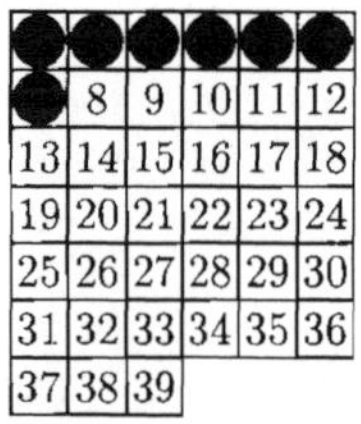

Finnish Lotto 7/39
August 27, 1994

Figure 5.1 Most popular number combinations from Tables 5.1–5.4.

Since you presumably don't want to share a jackpot with thousands of other winners, you should observe the following *first law of foolish combinations,* which is subordinate to the categorical imperative for lotto players:

Never choose a combination of numbers using an arithmetic progression.

The following four examples illustrate the consequences of acting against this principle. For the Finnish Lotto 7/39, matching six of the seven winning numbers (without matching the bonus number) means winning a division 3 prize. In 1991, the average division 3 prize was 4,600 Finnish marks, with a maximum of 10,233 Finnish marks. However, for drawing 47 of 1991, each of the division 3 prize winners had to be content with only 645 Finnish marks. (By the way, this was the smallest division 3 prize since the beginning of the Finnish Lotto 7/39 in 1990!) The simple explanation for this curious fact is that the outcome of the 47th drawing was the nearly perfect arithmetic progression 15, 32, 33, 34, 35, 36, 37.

As a second example, consider the New Zealand Lotto 6/40, which conducted its first drawing on August 1, 1987. In this lotto, matching five of the six winning numbers means winning a division 3 prize. With the notable exception of drawing 320 on September 11, 1993, all division 3 prizes in 1993 were in the range between \$421 and \$829 (New Zealand dollars). However, due to the nearly arithmetic progression outcome 31, 32, 33, 35, 36, 38, the division 3 prize payout for drawing 320 was only \$179.

A third example is provided by the nearly arithmetic progression 8, 11, 18, 25, 32, 39 which was drawn on September 9, 1991, in the Indiana 6/44 Lotto Cash. This combination resulted in a minimum match 5 payoff of only \$561 U.S. for the year 1991, compared with an average match 5 payoff of \$1,375.

As a final example, consider the combination 11, 14, 15, 17, 18, 19, which was drawn on November 14, 1992, in the California Super Lotto 6/51. It resulted in the minimum match 5 payoff of only \$684 U.S. for the year 1992. In comparison, 75% of all draws in 1992 had a division 3 (5/6) payout of more than \$1,500, with a maximum prize of \$2,952.

5.2 Choosing Winning Combinations of Previous Draws

Take a second look at Tables 5.1 through 5.4 and examine the extremely popular but apparently irregular combinations 1, 4, 15, 22, 26, 46 (3rd position in Table 5.1); 5, 27, 34, 35, 37, 49 (3rd position in Table 5.2); 8, 14, 17, 25, 31, 36 (4th position in Table 5.3); and 11, 15, 18, 30, 34, 35, 36 (6th position in Table 5.4). The solution to this mystery is very simple: Each of these four combinations is the outcome of the respective previous draw. Moreover, the tenth most frequent combinations in Tables 5.1 and 5.4 are the outcomes before that draw. So, for whatever reason, choosing winning combinations of past drawings seems to be a very popular strategy.

To investigate the effect of copying past drawings' winning numbers in particular (and the most foolish player behavior in general), consult Riedwyl (1990). This book lists each combination that was chosen at least 50 times for the February 10, 1990, drawing of the Swiss Lotto 6/45. Since the total number of combinations investigated by Riedwyl was 16,862,596, and the number of all combinations in a 6/45 lottery is 8,145,060, each combination listed was chosen at least 25 times above average.[1]

To illustrate the effect of copying winning numbers from past drawings, Figure 5.2 shows (on a logarithmic scale) the number of combinations chosen for drawing 34 of 1994 for the Finnish Lotto 7/39 that are copies of the winning numbers of the jth preceding week. It is striking to see that, even after a time lag of two years, a winning combination is still popular enough to be chosen about 100 times.

Since the same phenomenon was observed in Riedwyl (1990, p. 53) for the Swiss 6/45 Lottery and in Henze (1997, p.157) for the German Lotto 6/49, it would seem that any winning combination makes a lasting impression on players. A possible explanation for

[1]Hans Riedwyl is very much indebted to the Swiss Lottery Corporation for making these data available to him.

the popularity of past winning numbers might be that players who win division 2 or 3 prizes consider their particular winning combinations lucky and adopt them for future draws.

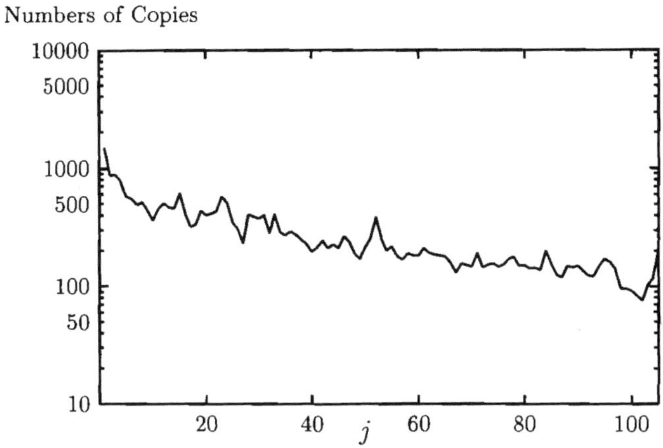

Figure 5.2 Number of combinations representing copies of the winning numbers of the jth preceding week for the August 27, 1994, draw of the Finnish Lotto 7/39.

Summarizing, this is the *second law of foolish combinations*:

Never choose a combination of numbers from winning combinations of previous draws.

To avoid making this error, as an informed lotto player you should keep track of winning numbers on a weekly basis. In particular, you should discard any of your own division 1, 2 or 3 winning combinations and not use them for a long time. (This might be hard for some of you!)

To illustrate the disappointing consequences of copying winning combinations from previous draws, consider the German

Lotto 6/49. This winning combination—**4**, 18, **21**, **25**, **26**, **31**—of March 11, 1995, was nearly repeated only seven weeks later on April 29, 1995, when the numbers **4**, 13, **21**, **25**, **26**, **31** showed up. Given the lasting effect that each winning combination has, an unusually large number of players would likely match five of the six winning numbers of April 29, 1995. Indeed, there were 7,326 division 4 prize (5/6–) winners that week, compared with the usual range of 1,200 to 2,500 winners. However, given the parimutuel payout system, each of these 7,326 winners received only 1,773 DM, compared with the usual division 4 prize of between 5,000 and 11,000 DM.

The *only* reason for staying away from combinations of previous draws is the apparent popularity of these combinations among other players. Since the drawing device doesn't care about the results of past drawings, each combination—regardless of whether it has already shown up or not—continues to have the same chance of winning the jackpot. Thus, a sentence like this (taken from an advertisement for a lottery book)—"Statistically, every combination has the same chance of winning the jackpot—however once a set of numbers has been a winner, history shows that it is extremely unlikely that it will ever win again!"—may give you the misleading impression that once a set of numbers has been drawn, it might have a smaller probability of winning the jackpot in future drawings. This is downright nonsense! Moreover, since there is only a fixed number of possible winning combinations, there must be a first repetition of some previously drawn set of numbers at some point in the future.

For instance, there are 13,983,816 possible combinations in a 6/49 lottery. How many draws will it take, on average, until the first repetition of a previously drawn set of numbers? Maybe you will be astonished to hear that this average number is only 4,687 draws. Yet keep in mind that 4,687 draws will take 45 years, if there are 2 draws per week. That concept should give you a feeling for the relatively short histories of existing lotteries. By the way, the first repetition of a winning combination in the German Lotto 6/49 occurred on June 21, 1995, after 3,016 draws, if the winning combinations of the two draws taking place on Wednesdays and the draws conducted on Saturdays are pooled. (For more details on how long it will take for the first repetition of some previously drawn combination for a general r/s lottery, see Chapter 8, Section 8.4.)

5.3 Choosing "Hot" and "Cold" Numbers

A feature common to all number lotteries is that some numbers come up more often than others. As a data set, consider that of the Canadian Lotto 6/49, which started on June 12, 1982, and thus has a fairly long history. Figure 5.3 shows the frequencies of winning numbers (excluding bonus numbers) after draw 1,220 of September 30, 1995.

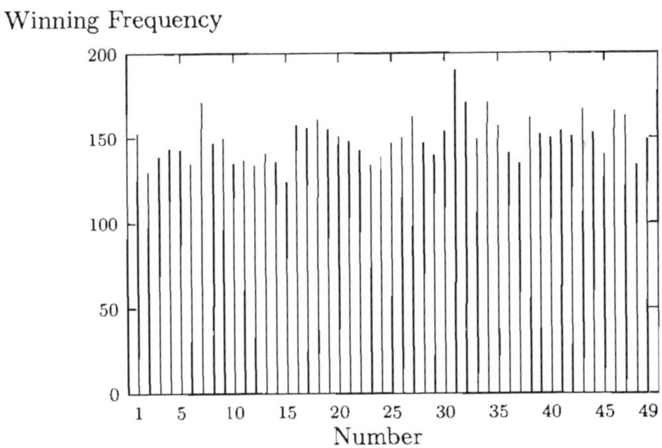

Figure 5.3 Frequency chart of winning numbers (excluding bonus numbers) after 1,220 drawings of the Canadian Lotto 6/49, September 30, 1995.

One of the greatest misconceptions of many lotto players is that the observed variation of the winning frequencies of past drawings should help predict the outcomes of future drawings. For example, one group of players always chooses the six most frequently selected ("hot") numbers, hoping these numbers will keep appearing. Thus, for draw 1,221 of the Canadian Lotto, those players would choose the combination 7 (171), 31 (190), 32 (171), 34 (171), 43 (167), 46 (166), where the numbers in parentheses are the present winning frequencies. There is also the contrary opinion that if a number hasn't come up often, it's "due". Followers of this theory always select the six least frequently hit ("cold") numbers. They would choose the numbers 2 (130), 12 (134), 15 (124), 23 (134), and 48 (134) plus one of the numbers 6 (135), 10 (135), or 37 (135) for the 1,221st draw.

Picking either the six hottest or coldest numbers is an extremely foolish style of play. Again, the drawing device operates independently of past drawings; it doesn't care about favored or neglected numbers. The fact that some numbers come up more often than others is simply *natural chance variation.* This means that the observed winning frequencies—although covering the presumably large range from 124 (for the number 15) to 190 (for the number 31)—are nevertheless in accordance with the conception of a "drawing device without memory" and the idea that each combination has the same chance of being drawn. A check of a mathematical model that assumes an equiprobability of lotto numbers may be done by performing the chi square test (see Chapter 8, Section 8.3).

Interestingly, many people believe that the *range* of the winning frequencies —that is, the difference between the largest and the smallest of these frequencies—after n draws should probably become smaller and smaller as the number n of draws increases. This widespread misconception suggests that in the long run, there should be some sort of equalization of winning frequencies, which attributes a memory to the drawing device. However, as a consequence of statistical independence of the results of weekly draws and an equal chance of winning for each combination, the random fluctuations of the range of winning frequencies must grow with the square root of the number of draws (see Chapter 8, Section 8.2). So, although you might think that there are very hot and very cold numbers, which probably get even hotter or colder over time, you aren't observing anything more than normal statistical variation.

To illustrate how popular hot and cold number strategies are, consider that for the draw of October 16, 1993, in the German Lotto of Baden–Württemberg, 393 players followed the "play the hottest six" strategy, and 192 players followed the "pick the coldest six" strategy (from a total of 6,803,090 combinations). For the draw of February 10, 1990, of the Swiss Lotto 6/45, 992 players selected the six hottest numbers (with a total of 16,862,596 combinations chosen).

These examples illustrate the *third law of foolish combinations*:

Never pick the hottest or coldest numbers.

Note, as well, that you also go against the categorical imperative for lotto players if you choose a combination of the hottest or coldest numbers. Since there are only 400 (= 20 × 20) ways of choosing three numbers from the hottest six group and three from the coldest six group, you are in great danger of sharing the same combination with many other players when you follow a hot or cold number strategy.

5.4 Choosing Outstanding Numbers

A further widespread misconception among many lotto players is that certain numbers are due to come up if they have not been picked for a long time. To encourage this erroneous belief (and to facilitate a suitable playing strategy), so-called drawings-since-hit tables are published weekly, indicating how many drawings have taken place since each number was last selected. Table 5.5 shows a drawings-since-hit table for the 7/35 Swedish Lotto "1" after the draw of July 15, 1995. In looking at the Table, notice that there are seven 0 entries corresponding to the previous draw's winning numbers: 7, 12, 16, 20, 22, 27, 28. On the other hand, some numbers have not been drawn for weeks, including 11 (15 weeks), 21 (11 weeks), 4 and 19 (each 10 weeks), 10 (8 weeks), and 6, 14, and 23 (7 weeks).

1	**2**	**3**	**4**	**5**	**6**	**7**
2	1	6	10	1	7	0
8	**9**	**10**	**11**	**12**	**13**	**14**
4	1	8	15	0	6	7
15	**16**	**17**	**18**	**19**	**20**	**21**
3	0	6	3	10	0	11
22	**23**	**24**	**25**	**26**	**27**	**28**
0	7	4	1	4	0	0
29	**30**	**31**	**32**	**33**	**34**	**35**
4	1	2	2	3	1	1

Table 5.5 Drawings-since-hit table for the Swedish Lotto 7/35 after the draw of July 15, 1995.

Contrary to the typical player's fallacy, these presumably long waiting times are absolutely normal. As noted earlier, pure chance produces a random fluctuation of winning frequencies and thus a random fluctuation of waiting times as well (see Chapter 8, Sections 8.5 and 8.6). Moreover, even if some number didn't come up for, say, 100 weeks or more, this would not have any consequence for future drawings.

However, some players hold the misconception that choosing the six most outstanding numbers will increase their chances of winning. Followers of this theory would choose the numbers 11, 21, 4, 19, and 10 plus two of the numbers 6, 14, and 23 for the next draw of the Swedish Lotto 7/35 (that is, on July 22, 1995). But since "Choose the most outstanding numbers" is a rule, it is against the categorical imperative for lotto players. Namely, if you play according to this simple strategy, you risk having to share the grand prize with many other players, should you win. To see how popular this strategy is, note that for the draw of February 10, 1990, of the Swiss Lotto 6/45, 1,168 players followed the "Choose the most outstanding numbers" strategy. The corresponding number of players for the German Lotto 6/49 (October 16, 1993) was 460.

Interestingly, the ratio

$$1,168 \;/\; 460 \;\approx\; 2.54$$

is nearly the same as the ratio

$$16,862,596 \;/\; 6,803,090 \;\approx\; 2.48,$$

which reflect the numbers of investigated combinations for the Swiss and German Lottos, respectively.

Since the behavior of lotto players seems very consistent all over the world, you should assume that the most outstanding number combinations are played approximately 700 times per 10,000,000 chosen combinations. Thus, the *fourth law of foolish combinations* is

Never pick the most outstanding numbers.

5.5 Choosing Winning Numbers from Other Countries or States

Yet another popular strategy among players is to pick the winning numbers of previous draws from the number lotteries of other countries or states. Take a good look at Table 5.6, which reveals the winning combinations (and certain modifications thereof) from other countries that players chose most frequently for the February 10, 1990, draw of the Swiss Lotto 6/45. Interestingly, players from Switzerland copied previous winning combinations from lotteries in countries such as Belgium, Finland, Italy, and even Australia. In column 5 of the Table, S denotes the supplemental or bonus number. Thus, Austria's winning combination of February 4, 1990, was 15, 16, 20, 24, 25, 43, with the bonus number 30;

Country	Draw	Date	Combinations/ Modifications						S	Frequency
Austria	–	02/04/90	15	16	20	24	25	43	30	1,283
			15	16	20	24	25	30		81
Germany	A	02/07/90	10	12	33	36	38	47	32	–
			10	12	33	36	38	45		1,288
			10	12	33	36	38	32		1,739
	B	02/07/90	8	10	27	44	48	49	9	–
			8	10	27	44	43	45		155
			8	10	27	44	9	45		218
	–	02/03/90	20	29	35	42	46	48	45	–
			20	29	35	42	44	45		186
			20	29	35	42	43	45		146
	A	01/31/90	1	15	26	33	41	43	29	613
	B	01/31/90	2	4	5	14	22	31	44	553
	–	01/27/90	6	9	10	12	37	42	44	752
France	–	02/07/90	4	8	21	23	42	49	37	–
			4	8	21	23	42	45		2,722
			4	8	21	23	42	44		106
			4	8	21	23	42	43		94
			4	8	21	23	42	37		679
	–	02/03/90	11	31	36	44	45	48	2	–
			11	31	36	44	45	2		577
	–	01/31/90	3	8	14	20	25	28	7	824

Table 5.6 Frequencies of winning combinations (and modifications thereof) from other countries' lotteries chosen for the February 10, 1990, draw of the Swiss Lotto 6/45.

1,283 players picked this combination for the Swiss Lotto draw six days later. In addition, 81 players replaced the largest number 43 by the bonus number 30 and thus chose the combination 15, 16, 20, 24, 25, 30. (Recall from page 42 that these frequencies correspond to a total of 16,862,596 combinations played for the February 10, 1990, draw of the Swiss Lotto 6/45.)

As for the winning combinations from Germany, you should know that there are two different 6/49 lotteries: the traditional one that started in 1955 and has draws on Saturdays, and another one that started in 1986 and has two draws (called A and B) on Wednesdays. Note that because of the number 47, the combination 10, 12, 33, 36, 38, 47, which came up in draw A of February 7, 1990, could not be chosen for the draw of the Swiss Lotto 6/45 three days later. Even so, 1,288 players simply replaced 47 by the maximum admissible number 45, and another 1,739 players replaced 47 by the bonus number 32, therefore choosing the combination 10, 12, 32, 33, 36, 38 for the Swiss Lotto draw of February 10, 1990. A similar phenomenon was observed for the German Lotto 6/49. For the draw of October 16, 1993, 191 players (out of a total of 6,803,090) chose the previous winning numbers from the Swiss lottery, and the French and Austrian winning combinations of October 9, 1990, were also chosen over 50 times each—more than *100 times more often than average.*

The strategy "Play the previous winning combination from the lottery of another country or another state" is a simple rule and seems fairly popular among players. As such, you should observe the *fifth law of foolish combinations*:

Never choose a previous winning combination from another country's or state's lotto.

As a final warning against choosing winning numbers from other countries'/states' lotteries, consider that for the German Lotto draw of June 18, 1977, 205 players (which is about 30 times more than average) chose the previous winning combination from the Netherlands Lotto 6/49, and call it luck or misfortune, but those numbers did indeed come up! After the jackpot was split among the winners, each person went away with the fairly moderate prize of 30,737.80 DM (then, approximately $12,800 U.S.).

5.6 Choosing Geometric Patterns

Do you choose your numbers according to geometric patterns? If you do, you will have good-looking tickets, but likely decrease your chances of winning big. As an illustrative warning, take a good look at the left-hand ticket in Figure 5.4, which shows the winning combination 24, 25, 26, 30, 31, 32 of the January 23, 1988, draw of the German Lotto 6/49, together with the bonus number 33. There were 222 first-prize winners (of a total of 150,931,641 played combinations), which was about 21 times above average. Furthermore, 1,509 players matched five of the winning numbers plus the bonus number, which was 23 times above average. The right-hand ticket, in which the numbers form a horseshoe (meant to bring lottery luck) or the letter U, was the winning combination of October 4, 1997. It was matched by 134 players, which was about 19 times above average.

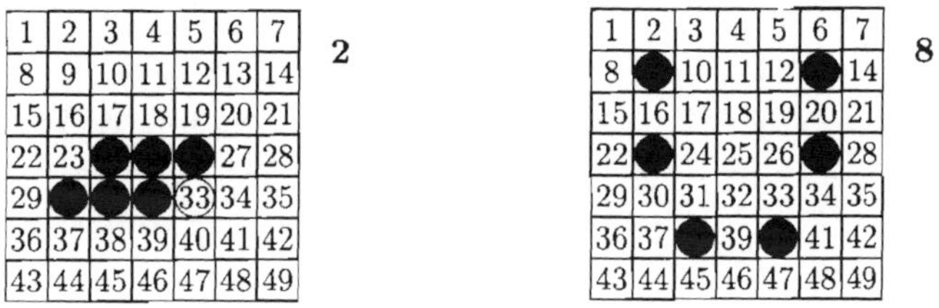

Figure 5.4 Popular geometric patterns of winning combinations for two draws of the German Lotto 6/49.

The number 2 appearing at the upper-right side of the left-hand game board is the *arithmetic complexity* of the combination 24, 25, 26, 30, 31, 32. For a general r/s lottery, the arithmetic complexity (denoted by AC) of a combination is defined as the number of positive differences between any two numbers in a combination, minus $(r - 1)$. The combination 24, 25, 26, 30, 31, 32 has an AC value of 2 $(= 7 - 5)$ because there are seven such differences: namely, 1 $(= 25 - 24 = 26 - 25 = 31 - 30 = 32 - 31)$, 2 $= (26 - 24 = 32 - 30)$, 6 $(= 30 - 24 = 31 - 25 =$

32 − 26), 7 (= 31 − 24 = 32 − 25), 8 (= 32 − 24), 5 (= 30 − 25 = 31 − 26), and 4 (= 30 − 26).

Likewise, the number 8 at the upper-right side of the right-hand game board is the AC value of the combination 9, 13, 23, 27, 38, 40. From the definition of arithmetic complexity, it follows that an arithmetic progression has the lowest possible AC value of 0. Moreover, the maximum possible AC value is 10 for a $6/s$ lottery (see Chapter 8, Section 8.9).

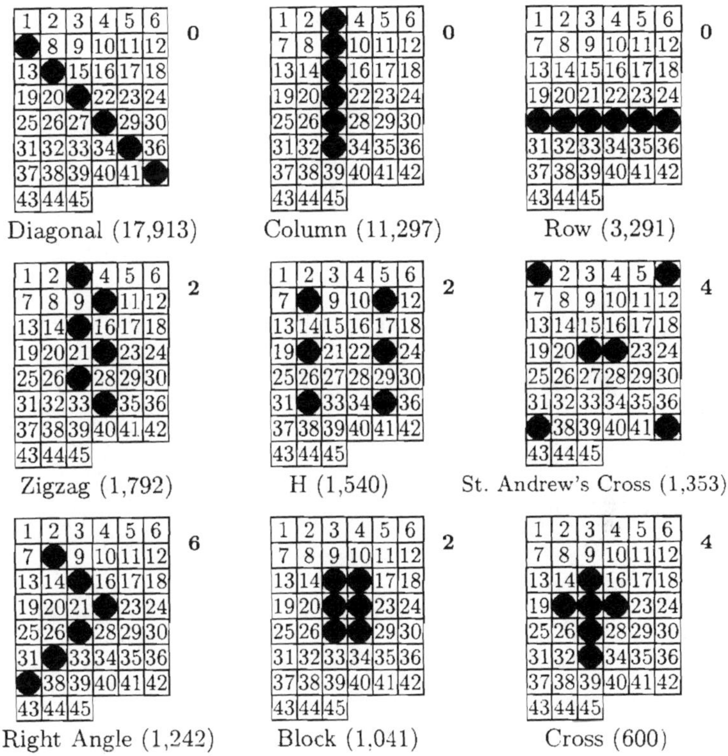

Figure 5.5 Popular geometric patterns and frequencies (numbers in parentheses) for the February 10, 1990, draw of the Swiss Lotto 6/45.

To understand the popularity of players' picking geometric patterns, Figure 5.5 shows how often several types of patterns were chosen for the Swiss Lotto 6/45 draw on February 10, 1990. As you can see, people chose not only diagonals, columns, and rows (all of which are arithmetic progressions that you already know are very popular) but also zigzags, Hs, right angles, blocks, and crosses. To assess how frequently these patterns are selected, remember that dividing each frequency by two yields the number of times *above average* each ticket was chosen. Note that with the exception of the "Right Angle," all of the popular geometric patterns have low AC values (4, at most).

Some more popular geometric patterns, which players chose for a German Lotto draw, are shown together with their AC values in Figure 5.6.

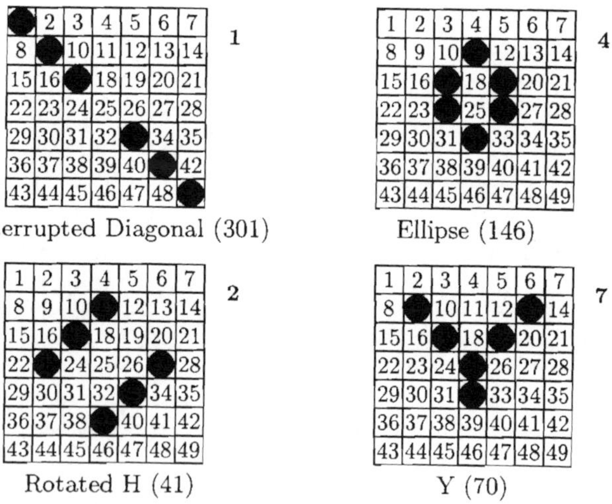

Figure 5.6 More popular geometric patterns and frequencies (numbers in parentheses) for the October 16, 1993, draw of the German Lotto in Baden–Württemberg.

On a 7 × 7 square game board such as this one, it might be tempting to choose a combination that is symmetric with respect to either the center (the two left-hand tickets in Figure 5.6) or the middle column (the two right-hand tickets). Recall that doubling the frequency of each ticket gives the number of times above average it was chosen.

These examples should convince you of the *sixth law of foolish combinations*:

> **Never choose numbers that form a geometric pattern or symmetric combination.**

5.7 Modifying Previous Winning Combinations

You learned in Section 5.2 that it's extremely popular—and thus foolish—to choose winning combinations of previous draws. But what about starting with last week's winning combination and modifying it in some way, so as to get a new set of numbers? Would the resulting combination likely be unique?

Not at all, as a simple example will demonstrate. Start with the winning combination $8, 14, 17, 25, 31, 36$ from the February 3, 1990, draw of the Swiss Lotto 6/45. This combination was chosen 12,008 times (out of a total of 16,862,596) in the subsequent drawing of February 10, 1990; it was the fourth most popular combination in that draw (see Table 5.2). Using a very simple rule—namely, "Add 1 to each number"—the modified combination $9, 15, 18, 26, 32, 37$ is obtained. However, even this modified combination was still popular enough to be chosen 2,342 times, which was more than 1,000 times above average! Similarly, the combination $7, 13, 16, 24, 30, 35$, formed by subtracting 1 from each number, was chosen 1,623 times in the drawing of February 10, 1990.

Interestingly, not only the "Add or subtract 1" rules are very popular. People also like to perform simple arithmetic operations like "Add 2 (or 3, 4, . . .) to each number" or "Subtract 2 (or 3, 4, . . .) from each number" of the winning combination from the last draw. Table 5.7 shows the frequencies of arithmetic modifications of the winning combination $8, 14, 17, 25, 31, 36$, chosen for the February 10, 1990, draw of the Swiss Lotto 6/45 out of a total of 16,862,596 lottery tickets. The modification "Add k to each winning number" is denoted by $+k$, and the modification $-k$ means "Subtract k from each winning number."

Combination						Modification	Frequency
8	14	17	25	31	36	–	12,008
9	15	18	26	32	37	+1	2,342
7	13	16	24	30	35	−1	1,623
10	16	19	27	33	38	+2	405
6	12	15	23	29	34	−2	306
11	17	20	28	34	39	+3	210
5	11	14	22	28	33	−3	160
12	18	21	29	35	40	+4	125
4	10	13	21	27	32	−4	137
13	19	22	30	36	41	+5	118
3	9	12	20	26	31	−5	110
14	20	23	31	37	42	+6	156
2	8	11	19	25	30	−6	160
15	21	24	32	38	43	+7	64
1	7	10	18	24	29	−7	61
16	22	25	33	39	44	+8	48
17	23	26	34	40	45	+9	48

Table 5.7 Frequencies of arithmetic modifications of the winning numbers from the February 3, 1990, draw chosen for the subsequent February 10, 1990, draw of the Swiss Lotto 6/45.

As you may have guessed by now, lotto players not only like to modify last week's winning numbers but also combinations from many weeks ago. Table 5.8 lists the frequencies of arithmetic modifications of the winning numbers of the jth preceding week, chosen for the draw of February 10, 1990, for the Swiss Lotto.

j	Modification	Frequency
2	+1	283
2	−1	257
2	−2	38
3	+1	205
3	−1	125
3	+2	28
4	+1	85
4	−1	70
5	+1	90
6	+1	33
6	−1	40

Table 5.8 Frequencies of arithmetic modifications of the winning numbers of the jth preceding week, chosen for the February 10, 1990 draw of the Swiss Lotto 6/45.

This players' strategy of modifying previous winning combinations by simple arithmetic operations has also been observed in the German and the Finnish Lottos. Players have used strategies involving *geometric modifications* of previous winning combinations as well. One example of a geometric modification is the *row shift* of a set of numbers, as illustrated in Figure 5.7.

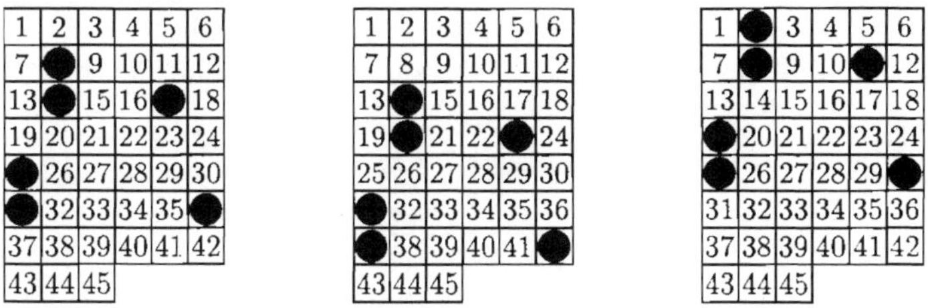

Previous Draw — Downward Shift — Upward Shift

Figure 5.7 Addition of the value 6 produces a downward shift; subtraction of the value 6 produces an upward shift.

The fact that row shifts are not only geometric but also arithmetic modifications of combinations (obtained by adding or subtracting the value 6) explains a curiosity of Table 5.7. Note that although the popularity of the arithmetic modifications $+k$ and $-k$ tends to decrease as k increases, the modification $+6$ (downward shift) is preferred to $+5$ and even to $+4$. Likewise, the modification -6 (upward shift) is preferred to -5 and -4.

1	2	3	4	●	6	7
8	9	10	11	12	13	14
15	16	17	18	19	20	21
22	23	24	25	26	●	28
29	30	31	32	33	●	●
36	●	38	39	40	41	42
43	44	45	46	47	48	●

Previous Draw (3,698)

1	2	●	4	5	6	7
8	9	10	11	12	13	14
15	16	17	18	19	20	21
22	●	24	25	26	27	28
●	●	31	32	33	34	35
36	37	38	39	40	●	42
●	44	45	46	47	48	49

Previous Draw, Reflected at Middle Column (101)

1	2	3	4	5	6	7
8	●	10	11	12	13	14
15	16	17	18	19	20	21
22	23	24	25	26	27	28
29	30	31	32	33	34	●
36	37	●	●	40	41	42
●	44	●	46	47	48	49

Previous Draw, Rotated about Center through 90° (34)

●	2	3	4	5	6	7
8	9	10	11	12	●	14
●	●	17	18	19	20	21
22	●	24	25	26	27	28
29	30	31	32	33	34	35
36	37	38	39	40	41	42
43	44	●	46	47	48	49

Previous Draw, Rotated about Center through 180° (124)

1	2	3	4	●	6	●
8	9	10	●	●	13	14
●	16	17	18	19	20	21
22	23	24	25	26	27	28
29	30	31	32	33	34	35
36	37	38	39	40	●	42
43	44	45	46	47	48	49

Previous Draw, Rotated about Center through 270° (32)

Figure 5.8 Geometric variants of a previous week's winning combination and frequencies for the October 16, 1993, draw of the German Lotto in Baden-Württemberg.

To provide more examples, Figure 5.8 shows some geometric modifications (together with their frequencies) of the previous week's winning combination $5, 27, 34, 35, 37, 49$ that were chosen (out of a total of 6,803,090 tickets) for the October 16, 1993, draw of the German Lotto 6/49 in Baden–Württemberg. As these tickets show, people like to perform *reflections* as well as *rotations* of combinations about various angles. Of course, a square game board—which has a center as well as two axes of symmetry—is ideally suited to these strategies.

These examples should suffice to convince you of the *seventh law of foolish combinations*:

> **Never choose a simple arithmetic or geometric modification of a previous winning combination.**

5.8 Playing with Numbers

This subsection illustrates the popularity of simple rules that are derived from the idea of *playing with numbers*. You already know from Section 5.1 that the most popular and thus most foolish strategy is to form an arithmetic progression. The following examples will show some more foolish approaches to playing lotto.

First, consider the "two numbers" rule. For instance, using only the numbers 1 and 2 for building a lotto combination, you obtain the combination $1, 2, 11, 12, 21, 22$. In the same way, you can use the numbers 1 and 3 to generate the combination $1, 3, 11, 13, 31, 33$. To assess the popularity of this simple strategy, look at Table 5.9, which lists the frequencies of all possible "two numbers" rules chosen for the February 10, 1990, draw of the Swiss Lotto 6/45. Since this lotto draw will be referred to many times in the following discussion, it will be abbreviated and called *the Swiss Lotto draw*.

Although these frequencies don't seem too impressive, recall that dividing each by 2 yields the number of times the combination was chosen above average.

Combination						Frequency	AC Value
1	2	11	12	21	22	90	2
1	3	11	13	31	33	41	5
1	4	11	14	41	44	23	5
2	3	22	23	32	33	46	5
2	4	22	24	42	44	19	2
3	4	33	34	43	44	17	5

Table 5.9 Frequencies and arithmetic complexities (AC values) of "two numbers" combinations chosen for the February 10, 1990, draw of the Swiss Lotto 6/45.

Another strategy derived from playing with numbers is "Start with 1, and always multiply by 2 to get the next number." The resulting combination $1, 2, 4, 8, 16, 32$ was chosen 193 times more often than average for the draw of October 16, 1993, of the German Lotto 6/49 (Baden–Württemberg) and 121 times more often than average for the Swiss Lotto draw. The rule "Start with 1, and always square to get the next number" seems a little less popular; even so, the resulting combination $1, 4, 9, 16, 25, 36$ of square numbers was still chosen 53 times more often than average for the above-mentioned German Lotto draw, and 58 times more often than average for the Swiss Lotto draw.

Some players choose a starting number a and another number d, where d stands for *difference*. The rule, then, is "Start with a, and add d to get the next number; then add $d + 1$ to get the next number; then add $d + 2$ to get the next number; and so on." For example, the combination $2, 5, 9, 14, 20, 27$ is obtained by starting with $a = 2$ and $d = 3$. Table 5.10 shows the frequencies and arithmetic complexities (AC) of some combinations formed by the "Always add one more number" rule that were chosen for the Swiss Lotto draw.

a	d	Combination						Frequency	AC Value
1	2	1	3	6	10	15	21	62	8
2	2	2	4	7	11	16	22	19	8
2	3	2	5	9	14	20	27	41	8
3	3	3	6	10	15	21	28	31	8
3	4	3	7	12	18	25	33	48	9

Table 5.10 Frequencies and arithmetic complexities (AC values) of some examples of the "Always add one more number rule," chosen for the February 10, 1990, draw of the Swiss Lotto 6/45.

You are generally in danger of forming a popular combination if you play sequences of numbers with certain nice properties—for example, the square numbers sequence $1, 4, 9, 16, 25, 36$ (see above). Two similarly nice sequences are those of prime numbers and Fibonacci numbers.

A *prime number* is any number greater than 1 that cannot be evenly divided by a whole number greater than 1. The first 10 prime numbers are $2, 3, 5, 7, 11, 13, 17, 19, 23$, and 29. To assess the popularity of a strategy based on using some sequence of consecutive prime numbers, examine Table 5.11. It lists the

Combination						Frequency	AC Value
2	3	5	7	11	13	44	5
3	5	7	11	13	17	119	2
5	7	11	13	17	19	108	2
7	11	13	17	19	23	162	2
11	13	17	19	23	29	53	3
13	17	19	23	29	31	45	4
17	19	23	29	31	37	40	4
19	23	29	31	37	41	23	4
23	29	31	37	41	43	67	4

Table 5.11 Frequencies and arithmetic complexities (AC values) of sequences of consecutive prime numbers chosen for the February 10, 1990, draw of the Swiss Lotto 6/45.

frequencies and arithmetic complexities (ACs) of all possible sequences of six consecutive prime numbers chosen for the Swiss Lotto draw.

Surprisingly, sequences of *Fibonacci numbers* seem to be fairly popular, too. There is a simple recursive scheme for generating Fibonacci numbers: Start with 1 and 2, the first two Fibonacci numbers, and add them to get the next Fibonacci number, which is therefore 3. Then add the last two Fibonacci numbers obtained (2 and 3) to get the next Fibonacci number, 5. Continuing in this way, you produce the first 10 Fibonacci numbers: $1, 2, 3, 5, 8, 13, 21, 34, 55, 89$. In the Swiss Lotto draw, the two consecutive sequences $1, 2, 3, 5, 8, 13$ and $3, 5, 8, 13, 21, 34$ of Fibonacci numbers were chosen 34 times and 82 times, respectively.

In view of the preceding examples, the *eighth law of foolish combinations* can be stated as follows:

Never choose a combination by playing with numbers.

5.9 The Too-Large-Numbers Dilemma

Sections 5.1 and 5.6 discussed the great attractiveness of playing arithmetic progressions and nice-looking geometric patterns, respectively. Indeed, each strategy for picking a set of numbers is so appealing that even when there is one number too large to choose (which will be explained in a moment), players try to *approximate* nice arithmetic or geometric patterns.

As an example, consider the arithmetic progression $1, 11, 21, 31, 41$, which is especially attractive because each number has the same last digit. However, in a $6/s$ lottery, you would have to select one more number. Given the arithmetic progression, that number, of course, should be 51; you might choose it for the California Lotto 6/51 or the New York Lotto 6/54. However, if you were playing a 6/49 Lotto or some other $6/s$ lottery in which s

is smaller than 51, you would be in a *too-large-numbers dilemma.* Table 5.12 shows the frequencies and arithmetic complexities (ACs) of some number combinations that resolved this dilemma; all were chosen for the October 16, 1993, draw of the Lotto 6/49 in Baden–Württemberg, Germany, and the Swiss Lotto draw. Recall that doubling the frequency of any combination from the German Lotto draw yields the number of times above average that combination was chosen; for the Swiss Lotto draw, halving the frequency of any combination will indicate how many times above average it was chosen.

Combination	Frequency (German Lotto)	Frequency (Swiss Lotto)	AC Value
1 11 21 31 41 **42**	124	310	4
1 11 21 31 41 **43**	114	205	4
1 11 21 31 41 **44**	93	299	4
1 11 21 31 41 **45**	109	700	4
1 11 21 31 41 **46**	76	–	4
1 11 21 31 41 **47**	43	–	4
1 11 21 31 41 **48**	50	–	4
1 11 21 31 41 **49**	378	–	4
1 **2** 11 21 31 41	62	151	4
1 11 **12** 21 31 41	32	86	4
1 11 21 **22** 31 41	56	143	4
1 11 21 31 **32** 41	26	64	4

Table 5.12 Combinations that resolved too-large-numbers dilemmas, as well as their frequencies and AC values, for the German Lotto 6/49, October 16, 1993 and Swiss Lotto 6/45, February 10, 1990.

Interestingly, the Florida State Lottery 6/49 paid a record minimum match 5/6 prize of only $441.50 on July 31st, 1993, when the combination $1, 14, 24, 34, 44, 46$ was drawn. This low payout is due to the fact that the first 5 of these numbers are part of a broken arithmetic progression. Imagine the jackpot winners' disappointment if, instead of 46, the number 4 would have been drawn!

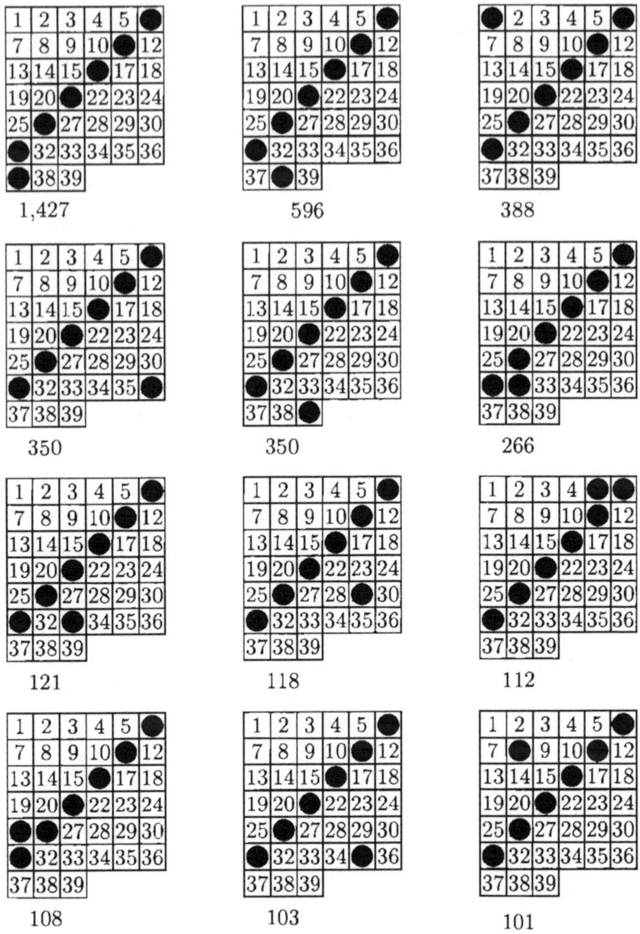

Figure 5.9 Combinations that resolved geometric too-much-numbers dilemmas and their frequencies for the August 27, 1994, draw of the Finnish Lotto.

Another too-large-numbers dilemma may occur if you try to select numbers according to aesthetic qualities. Looking back at the most popular combinations of Tables 5.1 through 5.4, notice that with the exception of the most popular Finnish combination, all were not only arithmetic progressions but also nice-looking upward diagonals. (There were even two of them in the case of the California Lotto; see Figure 5.1.)

So, you may wonder why the Finnish Lotto players would go against the trend and play an ugly favorite combination. Don't they like upward diagonals? They do, but they found themselves in a geometric dilemma. Trying to build an upward diagonal from the upper-right end to the lower-left end on a Finnish game board yields the six-number combination 6, 11, 16, 21, 26, 31. But there is still one number left to choose, because Finland has a 7/39 lottery. For players who are attracted by the aesthetic power of the diagonal 6, 11, 16, 21, 26, 31, there is only one way out of this too-large-numbers dilemma: Choose a seventh number that completes the combination in a way that approximates the desired geometric pattern as closely as possible. Figure 5.9 shows the most frequently chosen solutions to this dilemma. It is interesting to see that a great majority of players preferred to best approximate the upward diagonal by picking the number 37, which is as close to the ideal place as possible (that is, just left of it).

It is clear that the rule "Approximate an ideal arithmetic or geometric pattern" is subordinate to the categorical imperative for lotto players, and the examples given should convince you of the *ninth law of foolish combinations*:

Never try to approximate an ideal arithmetic or geometric pattern.

5.10 Foolish Play and Arithmetic Complexity

The example presented in this final section of the chapter serves two purposes: First, it shows that players' foolish behavior will likely be practically constant over time. Second, it demonstrates that the most foolish combinations can be classified into two categories: those based on aesthetic aspects, and those that copy past winning numbers.

Table 5.13 ranks the 35 most frequently (manually) chosen combinations for October 16, 1995 (that is, week 40 in 1995), draw of the German Lotto 6/49 in Baden-Württemberg.[2] The column labeled "Selection Above Average" indicates how many times above average each combination was chosen. In the column denoted "Rank", the number in parentheses gives the combination's rank in the previous week's draw (in player preference), whereas the number in brackets gives the rank of the combination exactly two years ago (that is, its rank in popularity for the draw of October 16, 1993). A [-] means that the combination was of normal (average) popularity.

In the column headed "Features", *PRO* means an arithmetic progression and a lowercase *pro* means a broken progression. *DIA*, *ROW* and *COL* stand for a diagonal, row, and column, respectively. *PAT* means a nice geometric pattern, and *SYM* means a combination that is symmetric with respect to the middle row, the middle column, or the center of the quadratic play slip.

Week j means that the combination is a copy of the winning numbers drawn in week number *j* in the year 1995, and *Week* 40*A* and *Week* 40*B* denote the winning numbers of the Wednesday Lotto Draw *A* and Draw *B* of week 40, respectively. Note that with the exception of all combinations that are copies from past drawings, the rankings of the regular foolish combinations are nearly unchanged over two years.

The final column shows the arithmetic complexity (AC) of each combination. It is striking to see that all of the nice-looking foolish combinations (that is, all those chosen for aesthetic purposes) in Table 5.13 have low arithmetic complexities (of at most 4),

[2]Norbert Henze would like to thank the lottery commission of the Federal State Baden-Württemberg of Germany for providing him with this data.

Rank			Combination						Selection Above Average	Features	AC Value
1	(1)	[9]	1	9	17	25	33	41	7,871	PRO, DIA	0
2	(2)	[1]	7	13	19	25	31	37	7,645	PRO, DIA	0
3	(4)	[2]	7	14	21	28	35	42	6,957	PRO, COL	0
4	(–)	[–]	6	11	14	15	25	46	6,667	Week 39	10
5	(5)	[4]	1	2	3	4	5	6	6,508	PRO, ROW	0
6	(6)	[5]	4	11	18	25	32	39	5,688	PRO, COL	0
7	(7)	[7]	6	12	18	24	30	36	4,770	PRO, DIA	0
8	(8)	[6]	13	19	25	31	37	43	4,617	PRO, DIA	0
9	(9)	[8]	9	17	25	33	41	49	4,065	PRO, DIA	0
10	(10)	[10]	8	16	24	32	40	48	3,818	PRO, DIA	0
11	( )	[-]	4	18	20	31	41	43	3,407	Week 40B	8
12	(3)	[-]	20	24	27	28	38	40	3,304	Week 38	9
13	(–)	[–]	1	4	6	12	32	39	3,294	Week 40A	10
14	(14)	[18]	1	10	20	30	40	49	3,045	pro	3
15	(15)	[16]	44	45	46	47	48	49	2,865	PRO, ROW	0
16	(17)	[15]	5	10	15	20	25	30	2,815	PRO, PAT	0
17	(16)	[14]	2	10	18	26	34	42	2,750	PRO, DIA	0
18	(18)	[21]	3	10	17	24	31	38	2,708	PRO, COL	0
19	(12)	[–]	19	21	29	37	40	43	2,576	Week 37	8
20	(20)	[24]	11	18	25	32	39	46	2,476	PRO, COL	0
21	(21)	[22]	10	18	24	30	33	44	2,432	unknown[4]	7
22	(19)	[20]	1	7	22	28	43	49	2,419	SYM, PAT	2
23	(23)	[23]	1	8	15	22	29	36	2,254	PRO, COL	0
24	(24)	[19]	43	44	45	46	47	48	2,166	PRO, ROW	0
25	(22)	[-]	7	16	23	37	41	42	2,077	Week 36	10
26	(25)	[ ]	6	11	25	32	37	42	2,033	Week 35	6
27	(26)	[25]	1	2	3	47	48	49	1,995	SYM	2
28	(28)	[28]	5	12	19	26	33	40	1,974	PRO, COL	0
29	(27)	[26]	14	20	26	32	38	44	1,755	PRO, DIA	0
30	(30)	[63]	2	9	16	23	30	37	1,706	PRO, COL	0
31	(31)	[29]	6	13	20	27	34	41	1,545	PRO, COL	0
32	(29)	[-]	13	17	20	27	33	48	1,497	Week 34	9
33	(32)	[36]	3	13	23	33	43	49	1,481	pro	4
34	(33)	[31]	8	15	22	29	36	43	1,478	PRO, COL	0
35	(35)	[32]	7	13	19	31	37	43	1,368	SYM	1

Table 5.13 Combinations that illustrate foolish play and arithmetic complexity for the October 16, 1995, draw of the German Lotto 6/49.

[4] The permanent popularity of this combination is an open problem.

whereas all of the foolish combinations obtained by copying winning numbers of past draws have high AC values (of at least 6). Since in a 6/49 lottery, more than 96% of all combinations have arithmetic complexities of at least 6 (see Chapter 8, Section 8.9), arithmetic complexity is a useful variable for discriminating between *simple combinations*, which exhibit much arithmetic or geometric regularity, and *really random combinations.*

Note, however, that some arithmetically regular combinations have high AC values, such as all combinations formed according to the "Always add one more number" rule (which have AC values of at least 8; see Table 5.10) or the combinations $1, 2, 4, 8, 16, 32$ (powers of 2) and $1, 4, 9, 16, 25, 36$ (square numbers), each of which has the highest possible AC value of 10. Moreover, not all geometrically appealing combinations have low AC values, such as the "Right Angle" in Figure 5.5 and the "Y" in Figure 5.6.

This chapter singled out the most popular and thus most foolish strategies for play. The overriding feature of foolish play is the application of simple rules in order to construct combinations in systematic ways. The problem is that such rules will likely have been used by many other players. Thus, when you play foolishly, your winnings (in the case of hitting the jackpot) will presumably be hundreds or even thousands of times smaller than average.

Chapter 6

Quick Pick: A Sensible Style of Play

Avoiding foolish combinations is easier said than done. There is a general danger connected with *manual play*, in particular (that is, when you choose your numbers yourself). Namely, for one reason or another, you have been attracted to some combination. But how can you assume that other players have not been drawn to the same combination? It seems that although hundreds of millions of lotto players all around the world go about their business separately, their behavior is much the same, and remarkably stable as well. For example, people often pick numbers that represent special occasions, such as birthdays and anniversaries, or they might pick the current day. Thus, if you were born on February 6, 1942, you might choose the numbers 2, 6, 19, 42; or on November 7, you (and probably countless others) might pick the numbers 7 and 11.

A suitable solution to the problem of having to select the numbers yourself is to let a computer do it by means of a random number generator. Such random number generators parade under various trademarked names, such as Quick Pick and Easy Pick. In an r/s lottery, a quick-pick machine (a computer that produces lotto numbers by means of some mathematical algorithm called lotto random number generator [see Chapter 8, Section 8.16]) selects r different numbers in the range 1 to s, so that each combination should have the same chance of occurring. Moreover, subsequent combinations of numbers should be statistically independent. Thus, a quick-pick machine tries to mimick the random selections of the drawing device.

Since quick-pick combinations selected by computer are actually chosen by some deterministic algorithm that is unknown to the public, they are not truly operating at random, and their quality must be assessed by statisticians. (There are several examples of bad random number generators!) The following discussion assumes that the quick-pick machine offered by your lottery corporation has passed such quality checks.

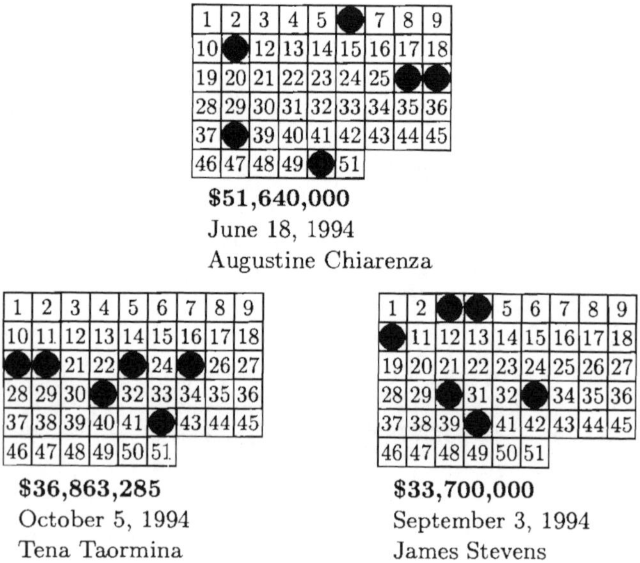

Figure 6.1 The three largest individual jackpot winners of the California Super Lotto 6/51, 1994.

To generate a quick-pick combination, you may use either a uniform random number generator (see Chapter 8, Section 8.16) or a deck of cards. In the latter method, simply take a deck of s cards, number them from 1 to s, and shuffle them thoroughly. Then draw r cards, and the corresponding numbers will form a quick-pick combination.

There is much evidence that selecting numbers using a quick pick gives high winnings in cases of hitting the jackpot. For instance, Augustine Chiarenza, a 72-year-old aerospace technician from Lompoc, became California's largest individual Super Lotto winner on June 18, 1994, when he won a \$51.6 million jackpot with a quick-pick combination (*Lottery Player's Magazine*, 1994, p. 4). Tena Taormina and James Stevens, the second- and the third-largest individual winners in 1994, also hit their jackpots with quick-pick combinations (see Figure 6.1).

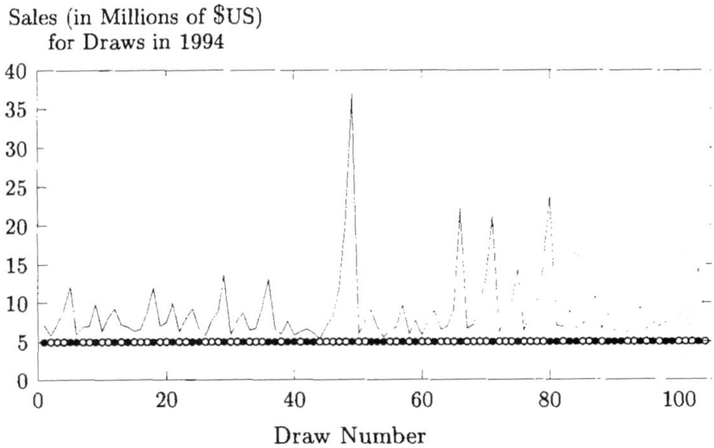

Figure 6.2 Ticket sales for the California Super Lotto 6/51, 1994.

Each of these extraordinary winnings occurred after at least four jackpot rollovers, as shown in Figure 6.2. The points show sales in millions of U.S. dollars for tickets in the California Super Lotto 6/51 during 1994. Draws in which the jackpot was hit are indicated with darkened circles ●; draws with jackpot rollovers are marked with open circles ○. Note the clear connection between jackpot rollovers and higher ticket sales.

Table 6.1 gives the number of jackpot winners for the California Lotto 6/51 in 1993 and 1994. As shown, the total number of jackpot hits due to quick-pick tickets in 1993 and 1994 was 62, which is about 63% of the total 98 jackpot hits. Interestingly, this percentage is nearly the same as that of quick-pick sales among total sales, which is largely constant over time and about 62% in 1997.

Jackpot Winners			
	1993	1994	Total
Frequency			
Manual Play	20	16	36
Quick-Pick Play	29	33	62
Mean Payoff			
Manual Play	$7,470,546	$7,466,040	$7,468,543
Quick-Pick Play	$9,013,138	$11,855,858	$10,526,198

Table 6.1 Number of jackpot winners in the California Super Lotto 6/51, 1993 and 1994.

These figures reflect the fact that playing quick pick does not increase the player's chance of hitting the jackpot. (The drawing device is completely impartial with respect to manual or quick-pick play!) However, when average payoffs are compared, quick-pick players typically win more than manual players. Over the two years 1993 and 1994, quick pick play increased the average payoff by about 40%, as illustrated by the last two rows of Table 6.1.

The discussion that follows addresses some consequences of quick-pick play. The main advantage of playing quick pick is that it protects players' choices from large-scale duplication. Furthermore, as long as there are foolish players and a substantial amount of manual play, playing quick pick may be an effective way of gaining an above-average prize, if you are enjoying one of those rare winning occasions.

It's important to note that this book would not have been written if all gambler selections were made completely at random. Without a certain percentage of foolish players, there is nobody to play against in a parimutuel payout system. A characteristic feature of quick-pick play is that the distribution of tickets chosen by the quick-pick machine is spread out more evenly over all possible combinations of numbers than under the play that actually occurs. As an example, use a data set (from Kadell and Ylvisaker, 1991) based on 111,221,666 combinations chosen for a draw of the California Lotto 6/53. (Recall from p. v that California had a 6/53 Lotto between June 1990 and December 1991.) Since there are 22,957,480 possible combinations in a 6/53 lottery, each single combination was selected

$$\frac{111,221,666}{22,957,480} \approx 4.8447$$

times *on average.*

If number combinations were selected completely at random, the proportion of combinations chosen k times would have, to an excellent approximation, a Poisson distribution with parameter 4.8447 (see Chapter 8, Section 8.13). Table 6.2 shows the frequencies of combinations bet k times, with k from 0 to 20, together with the corresponding *Poisson frequencies.* The latter are the expected numbers of combinations bet k times under complete random play, rounded to the nearest integer.

Note that the number of combinations not chosen by players is about 60% larger than the corresponding expected number of such combinations that would occur under quick-pick play. As a consequence, random (quick-pick) play is less likely to produce a jackpot rollover than the manual play that actually occurs. This fact is definitely not in the interest of the lottery commissions, because large carryovers lead to huge jackpots, excessive media attention, and high ticket sales.

k	Number of Combinations Bet k Times	Expected Value of Combinations Bet k Times under Quick-Pick Play
0	288,590	180,678
1	1,213,688	875,329
2	2,579,112	2,120,344
3	3,702,310	3,424,130
4	4,052,043	4,147,205
5	3,622,666	4,018,377
6	2,768,134	3,244,626
7	1,876,056	2,245,597
8	1,161,423	1,359,900
9	677,368	732,031
10	384,186	354,646
11	217,903	156,195
12	126,952	63,060
13	77,409	23,500
14	50,098	8,132
15	33,699	2,627
16	23,779	795
17	17,483	227
18	13,146	61
19	10,158	16
20	7,969	4

Table 6.2 Number of combinations bet k times and corresponding expected frequencies under quick-pick play for a California Lotto draw.

Also observe from Table 6.2 the "heavy tail" of the distribution of frequencies compared to that of the Poisson distribution with parameter 4.8447. This means that, near the bottom of Table 6.2, the numbers in the second column are much larger than their corresponding counterparts in the last column. Thus, the chance of having 20 winners under random play, for example, would be about 2,000 ($\approx 7{,}969/4$) times smaller than what occurs under actual play. In fact, strict quick-pick play would almost surely prevent 18 or more players from winning, if each combination were chosen 4.8847 times on average. A similar observation has been made by Riedwyl (1990).

Another consequence of strict quick-pick play is that the spread of the prize payouts would be much smaller than actually observed. To understand this point, imagine the extreme case that, for the draws of a lottery, each combination were chosen an equal number of times (let's say, 10 times); in that hypothetical case, the weekly payouts in each prize category wouldn't vary at all, because the number of winners would be known in advance. Under strict quick-pick play, there would be a fairly equal distribution of the bets over all combinations, so that the weekly payouts in each prize category would not show much statistical variation.

To assess the possible benefits gained from quick-pick play, consider this experiment: Ten data sets of 6/49 lottery tickets were generated by means of a computer, each set consisting of 5,000 randomly selected (quick-pick) combinations. Based on the information on stakes, payouts, and numbers of winners of previous draws for the Florida State Lottery 6/49, these 5,000 combinations were added to the tickets played for each draw during 1996 and 1995 and observed as *fictitious winnings* (that is, the winnings that these combinations would have produced if they had been played).

Table 6.3 reveals the fictitious average match 5/6 and match 4/6 payouts (in U.S. dollars), rounded to the nearest integer, for each set of combinations in the Florida State Lottery 6/49 during 1996 and 1995. For example, the fictitious average match 5/6 payoff for the first set of 5,000 combinations was \$1,970 in 1996 and \$1,510 in 1995, whereas the fourth set's average match 4/6 winnings in 1996 were \$65.

The reason for conducting several independent replications of one set of 5,000 combinations is to assess the statistical variation of the fictitious average payoffs. Note that due to the smaller number of winning tickets, this statistical variation over the ten sets of combinations is larger for the match 5/6 payouts than the match 4/6 payouts. The row denoted by "Average Payoff" contains the average payoffs evaluated over all ten sets of 5,000 combinations. These average prize amounts are always larger than the corresponding benchmark values indicated in the row "Manual Play Average Payoff". These represent the average

	1996	1995	1996	1995
Florida State Lottery	Match 5/6 Payoff		Match 4/6 Payoff	
Set 1	$1,970	$1,510	$64	$62
Set 2	$1,763	$1,466	$63	$63
Set 3	$1,570	$1,672	$64	$64
Set 4	$1,457	$1,429	$65	$64
Set 5	$977	$1,449	$65	$64
Set 6	$1,697	$1,577	$66	$63
Set 7	$1,565	$1,412	$63	$63
Set 8	$1,314	$1,741	$65	$64
Set 9	$1,528	$1,468	$63	$64
Set 10	$1,511	$1,711	$66	$63
Average Payoff	$1,530	$1,529	$64	$63
Manual Play Average Payoff	$1,376	$1,313	$62	$59
Performance	11.2%	16.5%	3.2%	6.8%

Table 6.3 Fictitious average payoffs for 10 sets of 5,000 quick-pick combinations and manual-play payoffs for the Florida State Lottery 6/49, 1996 and 1995 (in $U.S.).

payoffs realized under the manual play that actually occurred. Dividing the "Average" values by the corresponding "Manual Play" values yields the "Performance" figures shown in the last row. A performance of 16.5% means that the average match 5/6 value of $1,529 is 16.5% larger than the corresponding benchmark value of $1,313. As you can see, although the increase in average payoff is only slight for the match 4/6 prize amounts, it becomes more pronounced for the more interesting match 5/6 payoffs.

This experiment used data from the Florida State Lottery 6/49. Can the conclusions drawn from it about the benefits of playing quick pick be applied to lotteries in other countries? To test this, the same ten sets of 5,000 lottery tickets were added to the tickets played for each draw of the German Lotto 6/49 and the British National Lottery 6/49 in 1996 and 1995. Their fictitious average payouts, taken over all 50,000 combinations, are summarized in Tables 6.4 and 6.5, respectively.

	1996	1995	1996	1995
German Lotto	Match 5/6 Payoff		Match 4/6 Payoff	
Average Payoff	DM 7,792	DM 8,287	DM 138	DM 133
Manual Play Average Payoff	DM 7,530	DM 6,640	DM 133	DM 127
Performance	3.5%	24.8%	3.8%	4.7%

Table 6.4 Fictitious average payoffs for 10 sets of 5,000 quick-pick combinations and manual-play payoffs for the German Lotto 6/49, 1996 and 1995 (in deutsche Marks).

	1996	1995	1996	1995
British National Lottery	Match 5/6 Payoff		Match 4/6 Payoff	
Average Payoff	£1,772	£2,232	£68	£78
Manual Play Average Payoff	£1,575	£1,718	£63	£69
Performance	12.5%	30.0%	7.9%	13.0%

Table 6.5 Fictitious average payoffs for 10 sets of 5,000 quick-pick combinations and manual-play payoffs for the British National Lottery 6/49, 1996 and 1995 (in British pounds).

It is striking to see that in complete accordance with the data in Table 6.3, quick-pick play yields a notable and sometimes substantial increase in the average payouts for match 4/6 prizes and match 5/6 prize amounts.

The examples in this chapter should convince you that quick pick is a sensible style of play. If everyone played quick pick, there would never be very small jackpot winnings and there would rarely be very large jackpot winnings. Moreover, the spread of the payouts would be much smaller than occurs under manual play. In this sense, playing lotto would perhaps be boring under strict quick-pick play. However, thanks to the myriad of more or less foolish players, smart players can outperform strict quick-pick play with respect to long-term average payouts. Such intelligent styles of play will be studied in the next chapter.

Chapter 7

Intelligent Styles of Play

This is the chapter you've been waiting for! It will help you become a smart lotto player. But will doing so help you get rich?

To tell the truth, playing lotto is not bound to make you rich; it *may*, however. Given that only about 50% of the total lottery stake is returned to the winners (not to speak of taxes on the gains), lotto is an unfavorable game for the total group of players. And for individual lotto players, playing intelligently means, at best, selecting unpopular combinations in order to win more on those rare winning occasions and thus do better in the long run. Intelligent play is only possible because the manual betting that prevails leads to the overplaying many combinations and, consequently, the underplaying of many others. In this chapter, you will learn how to single out those underplayed combinations and assess the benefits of various strategies for intelligent play.

The following sections will give several recommendations for how to play lotto. Each will give you the freedom to choose from among *at least 10% of all possible combinations.* This point is very important, because with too specific advice—such as "Choose among the 15 most unpopular numbers only"—there will be too few combinations from among which to choose. So if many readers follow the advice, those combinations might quickly be overplayed.

In this context, the word *recommendation* is preferable and should be distinguished from the word *rule*, which was used in Chapter 5 to denote any specific strategy for choosing lotto numbers. The attractiveness of simple rules underlies most foolish styles of play, so always keep in mind the *categorical imperative for lotto players* : That is, never, *never*, choose any combination of numbers according to any simple rule. Following this imperative will at least prevent you from having to share a jackpot with many other players. (It is of course worse not to win at all, but getting $10,000 on a once-in-a-life time winning occasion, for example, and then seeing each of the next draw's jackpot winners receive $1,000,000 or even more—only for having matched an unpopular winning combination—is very disappointing.)

Quick pick is a sensible style of play, because it gives you a certain protection against large-scale duplication of numbers as well as above-average winnings in the long run (at least, as long as there are foolish players). However, you can do even better. Using the information on extreme payouts from Chapter 4 and that on foolish styles of play in Chapter 5, this chapter will investigate the influence of suitable variables (*factors*) on prize payouts. Moreover, based on such information, you will learn how to discard certain combinations (but not too many of them).

For several reasons, this investigation will study almost exclusively match 5/6 payoffs (without bonus numbers). First of all, matching five of the six winning numbers in a $6/s$ lottery is much more probable than winning the jackpot or matching five numbers plus the bonus number (see Tables 3.2 and 3.5). This means that a match 5/6 winning occasion is something that's really conceivable. In addition, matching five of the six winning numbers is much more profitable and thus much more interesting than merely matching four of six. In the second place, high match 5/6 payouts are likely to go along with high payouts in other prize categories; conversely, low match 5/6 payouts are likely to go along with low payouts in other categories (see Figures 4.12 through 4.14). Finally, there are always a lot of match 5/6 winners, so the application of statistical methods to this example is justified.

As a main data set, consider the match 5/6 payouts of the Florida State Lottery 6/49, evaluated over 201 consecutive draws from February 27, 1993 to December 28, 1996. Because of its exceptional character, the combination 1, 14, 24, 34, 44, 46 (see p. 63), which gave a record minimum match 5/6 payoff, was discarded from further study.

You should do the following investigations for the lottery in your state. All you need are data on winning numbers and prize payouts over some time span (at least one year) in which neither the lottery system nor the payout system changed. Much information is available from the World Wide Web by searching on the key words *lottery*, *lotto*, and *lotteries*.

7.1 Arithmetic Complexity

On page 52, the *arithmetic complexity (AC)* of a combination was defined as the number of different positive differences of its numbers, minus 5. In Section 5.10 (Chapter 5), you saw that many foolish combinations of numbers have low AC values, of at most 4.

Figure 7.1 shows a scatterplot of AC values (horizontal axis) against the natural logarithms of the match 5/6 prize amounts (vertical axis) for the 200 winning combinations of the Florida State Lottery, mentioned earlier. Note that 195 of the 200 combinations have AC values of at least 6, and the 4 combinations with ACs of 4 and 5 produced low payoffs. The lowest observed AC value of 3 belongs to the winning combination 30, 33, 36, 39, 44, 47 of January 21, 1995. Its fairly high payoff may be attributed to the fact that it does not contain any small numbers and only one number within the birthday range 1–31.

Figure 7.1 demonstrates that, as long as the AC value is at least 6, there is no notable dependence of the match 5/6 payout on arithmetic complexity. To avoid many foolish styles of play (see Tables 5.12 and 5.13), you should discard any combination that

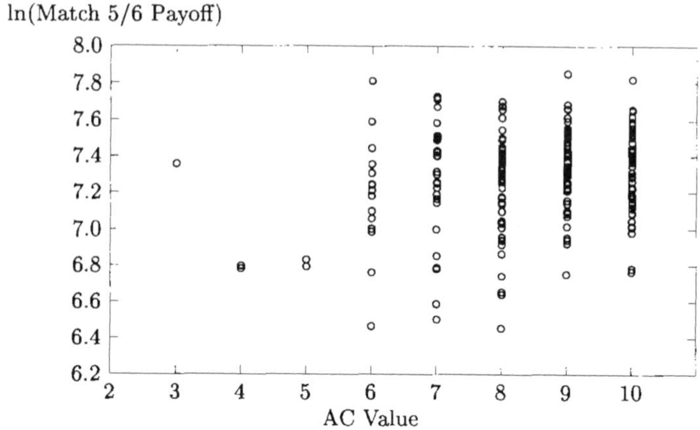

Figure 7.1 Scatterplot of arithmetic complexities versus ln(match 5/6 payoff) for the Florida State Lottery 6/49, 1993–1996.

has an arithmetic complexity of 5 or less. In a 6/49 lottery, this involves about 3.7% of all combinations (see Table 8.4). Discarding all combinations having AC values of five or less will be a *standing tacit assumption* that underlies all simulated average payoffs presented in this chapter.

7.2 Sums of Numbers

From Chapter 4, you know that a winning combination with a large sum of numbers is likely to go along with a high payoff (see Figures 4.10 and 4.11). Figure 7.2 gives a scatterplot of the natural logarithms of the sums of drawn numbers (horizontal axis) against the logarithms of the match 5/6 prize amounts (vertical axis) for the Florida State Lottery data set under discussion. The fairly high correlation shows a notable dependence of the ln(match 5/6 payoffs) on the influential variable ln(sum of drawn numbers).

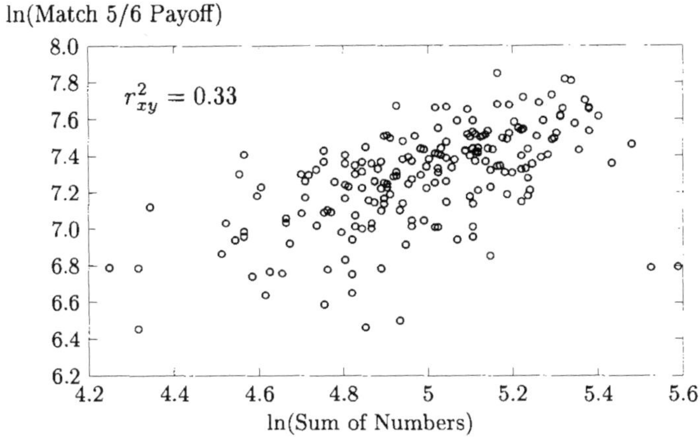

Figure 7.2 Scatterplot of the ln(sum of numbers) versus ln(match 5/6 payoff) for the Florida State Lottery 6/49, 1993–1996.

You should use the sum of numbers as a simple variable to distinguish between supposed popular and unpopular lottery tickets. Remember that many people choose birthdays or small numbers like 3, 7, and 9; thus, it is natural to avoid combinations having too many small numbers. This may be done by discarding combinations with too small sums of numbers.

In an r/s lottery, the sum of the numbers in a randomly chosen combination can be expected to be near $r \times (s+1)/2$ (see Chapter 8, Section 8.8). In a 6/49 lottery, this value is 150. Keeping this figure in mind, follow this *restricted quick-pick procedure*: Choose a combination of numbers at random, perhaps using a deck of cards (see p. 71). Play the combination if the sum of numbers is at least 174 and the AC value is at least 6 (recall the standing assumption!). Otherwise, discard the combination, shuffle the deck, and try again. For instance, the combination 2, 3, 13, 21, 24, 35 should be discarded, whereas the numbers 1, 23, 35, 40, 41, 48 should be accepted. (Of course, you should not play *this* combination, because many readers of this book might also do so.)

At first sight, the number 174 may seem somewhat arbitrary. However, using this figure will guarantee that you discard about

75% of all combinations in a 6/49 lottery (see Chapter 8, Section 8.8). Since there are exactly 3,230,773 combinations in a 6/49 lottery that satisfy both restrictions—the sum of numbers is at least 174 and the AC value is at least 6— this recommendation allows you to choose among 23.1% of all combinations.

To assess the benefits of following this recommendation, consider another experiment. Ten data sets of 6/49 lottery tickets were generated by means of a random number generator. Each set consisted of 5,000 combinations selected according to the restricted quick-pick procedure stated earlier. As done in Chapter 6, these 50,000 combinations were added to the tickets played for each draw of the Florida State Lottery 6/49, the German Lotto 6/49, and the British National Lottery 6/49 during the years 1996 and 1995. The winnings that these combinations would have earned, had they been played, were then observed.

The resulting fictitious average payouts (in $U.S.) for the Florida State Lottery are given in Table 7.1. The corresponding results for the German Lotto and British National Lottery are shown in Tables 7.2 and 7.3, respectively. Recall from p. 76 that each "Performance" figure in the last row of each table is obtained by dividing the simulated "Average Payoffs" by the "Manual Play Average Payoff" values (that is, the average payoff actually realized).

	1996	1995	1996	1995
Florida State Lottery	Match 5/6 Payoff		Match 4/6 Payoff	
Average Payoff	$1,800	$1,674	$73	$71
Manual Play Average Payoff	$1,376	$1,313	$62	$59
Performance	30.8%	27.5%	17.7%	20.3%

Table 7.1 Fictitious payoffs of 50,000 randomly chosen combinations (in which each sum of numbers is at least 174 and each AC value is at least 6) for the Florida State Lottery 6/49, 1996 and 1995 (in $U.S.).

A comparison of the values in Table 7.1 with the unrestrained quick-pick performance values given in Table 6.3 shows that this strategy of playing combinations with large sums of numbers and sufficiently high ACs yields a notable increase in the size of the average payoff, compared to the payoffs typical of strict quick-pick play, and a substantial increase when compared to actual play. The same conclusion can be drawn from a comparison of the values in Tables 7.2 and 7.3 with those in Tables 6.4 and 6.5, respectively.

	1996	1995	1996	1995
German Lotto	Match 5/6 Payoff		Match 4/6 Payoff	
Average Payoff	DM 8,875	DM 7,994	DM 146	DM 135
Manual Play Average Payoff	DM 7,530	DM 6,640	DM 133	DM 127
Performance	17.9%	20.4%	9.8%	6.3%

Table 7.2 Fictitious payoffs of 50,000 randomly chosen combinations (in which each sum of numbers is at least 174 and each AC value is at least 6), for the German Lotto 6/49, 1996 and 1995 (in deutsche Marks).

	1996	1995	1996	1995
British National Lottery	Match 5/6 Payoff		Match 4/6 Payoff	
Average Payoff	£2,600	£3,170	£82	£93
Manual Play Average Payoff	£1,575	£1,718	£63	£69
Performance	65.1%	84.5%	30.2%	34.8%

Table 7.3 Fictitious payoffs of 50,000 randomly chosen combinations (in which each sum of numbers is at least 174 and each AC value is at least 6), for the British National Lottery, 1996 and 1995 (in British pounds).

Although many of these performances are striking, you should *not* take the recommendation "Only play combinations with large sums of numbers" to absurd levels by playing combinations like

44, 45, 46, 47, 48, 49 and 43, 45, 46, 47, 48, 49, which have the largest and the second-largest possible sums of numbers, respectively. Since you know these combinations are foolish, don't play them. (Note, as well, that both combinations should be discarded because their AC values are too small.)

7.3 Edge Numbers

Suppose you are given a slip of paper containing the numbers 1, 2, 3, 4, arranged from left to right, and you are asked to circle one number at random. If you were to choose a number purely at random, the odds would be 1 in 4 for each number. Among a group of people, each number would be selected about 25% of the time. Many psychological experiments have shown, however, that the numbers on both ends are chosen far less often than should occur under pure chance.

In the same way, it should be expected that the numbers on the edge of a lottery game board are underplayed. To be precise, the *edge of a lottery play slip* is formed by the boxes of the top and bottom rows, the leftmost and rightmost columns, and the re-entrant corners (if any). Figure 7.3 shows the edges of two commonly used 6/49 lottery game panels.

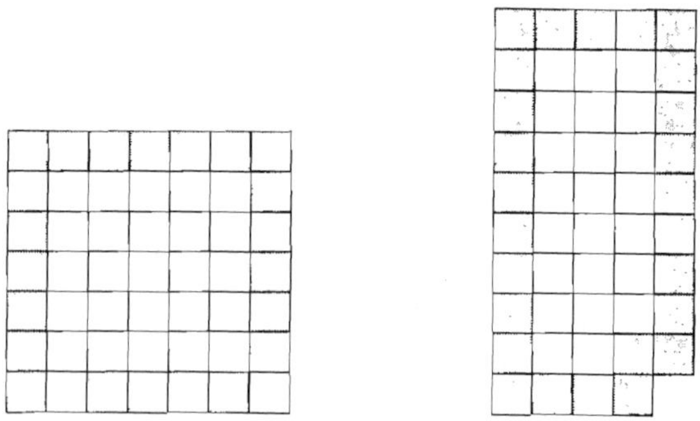

Figure 7.3 The edges of two typical 6/49 lottery game panels.

Since there are 24 edge boxes on the left-hand game board and 26 on the right-hand board, about half of the 49 numbers belong to the edge on each board. Thus, under quick-pick (random) play, it is expected that three of the six numbers are picked from the edge of the game panel. Table 7.4 shows the number of each combination in a 6/49 lottery whose edge number is k—that is, k of its numbers are on the edge of the game board ($k = 0, 1, . . ., 6$)—for each of the two game boards illustrated in Figure 7.3.

	Number of Combinations Having Edge Number k	
k	Board with 24 Edge Boxes	Board with 26 Edge Boxes
0	177,100	100,947
1	1,275,120	874,874
2	3,491,400	2,877,875
3	4,655,200	4,604,600
4	3,187,800	3,782,350
5	1,062,600	1,512,940
6	134,596	230,230
Total	13,983,816	13,983,816

Table 7.4 Distribution of edge numbers on a 6/49 lottery game board having 24 or 26 edge boxes.

The frequency distribution of the edge numbers for the combinations of the 200 draws of the Florida State Lottery under discussion is shown in Figure 7.4. Note that only two winning combinations have edge numbers of 0, and no combination has all six numbers on the edge of the game board. Moreover, most of the winning combinations have edge numbers of 2, 3, or 4. All of these findings are in complete accordance with the frequencies given in Table 7.4. More interesting with regard to intelligent play, however, are these findings: The two combinations having edge number 0 produced low match 5/6 payoffs, and the match 5/6 prize amount of a winning combination is likely to increase along with its edge number. Thus, it may be profitable to select combinations that have edge numbers of at least 4.

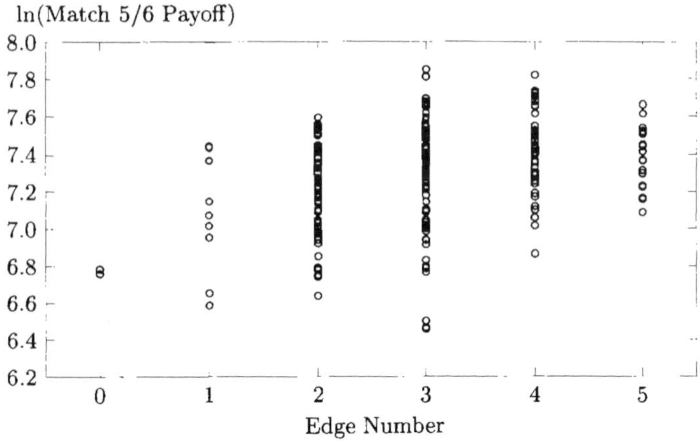

Figure 7.4 Scatterplot of edge numbers versus ln(match 5/6 payoff) for the Florida State Lottery 6/49, 1993–1996.

As seen before (in Chapter 6 and Section 7.2), the benefits of this strategy were assessed experimentally by simulating ten data sets of 6/49 lottery tickets with a computer. Each set consisted of 5,000 combinations having edge numbers of at least 4 and AC values of at least 6. All of the combinations were added fictitiously to the tickets played for each draw during 1996 and 1995 of the Florida State Lottery 6/49 and the British National Lottery 6/49, which have the same shape game panel and thus the same figures on the edge boxes.

	1996	1995	1996	1995
Florida State Lottery	Match 5/6 Payoff		Match 4/6 Payoff	
Average Payoff	$1,528	$1,481	$67	$66
Manual Play Average Payoff	$1,376	$1,313	$62	$59
Performance	11.0%	12.8%	8.1%	11.9%

Table 7.5 Fictitious payoffs of 50,000 randomly chosen combinations (in which each edge number is at least 4 and each AC value is at least 6), for the Florida State Lottery 6/49, 1996 and 1995 (in $U.S.).

	1996	1995	1996	1995
British National Lottery	Match 5/6 Payoff		Match 4/6 Payoff	
Average Payoff	£2,076	£2,670	£72	£86
Manual Play Average Payoff	£1,575	£1,718	£63	£69
Performance	31.8%	55.4%	14.3%	24.6%

Table 7.6 Fictitious payoffs of 50,000 randomly chosen combinations (in which each edge number is at least 4 and each AC value is at least 6), for the British National Lottery 6/49, 1996 and 1995 (in British pounds).

Tables 7.5 and 7.6 show the average payouts of all 50,000 tickets for the Florida State Lottery and the British National Lottery, respectively. It is interesting to compare the "Performance" data in Table 7.6 with those in Table 7.3. Although playing combinations with edge numbers of at least 4 yields striking increases in lottery payoffs compared to those won in manual play, you will perform significantly better if you choose combinations with large sums of numbers. Likewise, the "Performances" data in Table 7.1 are better than the corresponding values in Table 7.5. Thus, as a factor influencing payouts, the presence of numbers on the edge of the game board has some influence, but is less important than the sum of the numbers of a combination.

7.4 Cluster Numbers

Chapter 4 mentioned the phenomenon in which truly random choices of lotto numbers, when drawn on play slips, tend to have clusters of neighboring numbers. Since manual choices are not random but influenced by aesthetic points of view, among other things, it may be interesting to investigate the effects that certain kinds of clusters have on prize payouts.

To define the cluster number of a combination, consider adjacent, connected and isolated numbers. Any two numbers of a combination are called *adjacent* if their designated boxes on the play slip share a common edge or corner, as illustrated in Figure

Figure 7.5 Examples of adjacent lottery numbers.

7.5. Numbers are called *connected* if a path joins them via adjacent numbers. For example, in the rightmost combination in Figure 7.6, the numbers 24 and 35 are connected because a path joins 24 with the adjacent number 30 and 30 with the adjacent number 35. (Note that 24 is not adjacent to 35, however.) A number is *isolated* if no other number from the combination is adjacent to it. Thus, the middle combination in Figure 7.6 has the isolated numbers 1, 10, and 24.

The *cluster number* of a combination counts the number of isolated numbers, if any, plus the number of isolated groups of connected numbers, if any. As an example, look at the rightmost combination in Figure 7.6, in which there is one isolated number 26; one isolated group of the three connected numbers 24, 30, and 35; and one isolated group of the two connected numbers 42 and 47. The cluster number of this combination is therefore 3. The cluster number of the leftmost combination, which has four isolated numbers and one isolated group of two numbers, is 5, whereas the middle combination has a cluster number of 4.

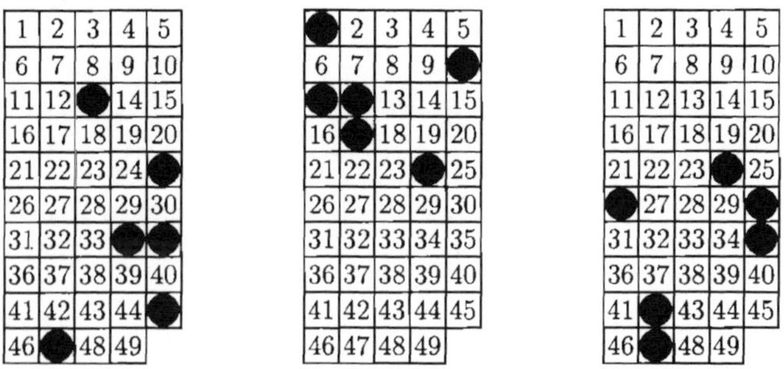

Figure 7.6 Combinations having cluster numbers 5, 4, and 3 (from left to right).

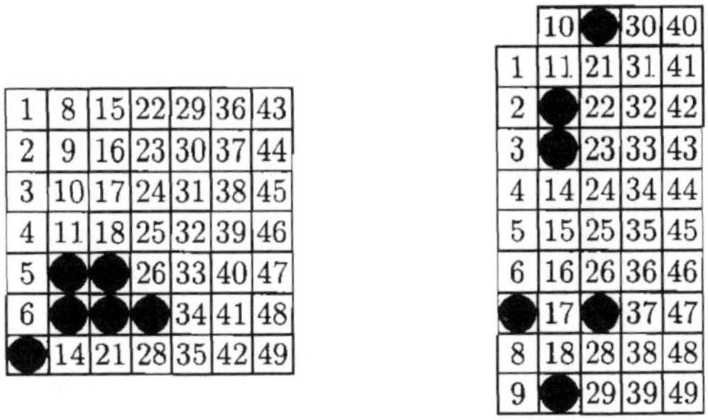

Game Board Used in Ontario

Game Board Used in All Other Provinces

Figure 7.7 Illustration of the dependence of the cluster number on the shape of game boards, Canadian Lotto, 1994.

Figure 7.7 illustrates that the cluster number of a combination may depend heavily on the shape of the game board. This fact will influence payouts in the Canadian Lotto, which uses a square game board in Ontario (left-hand board in Figure 7.7) and a rectangular game board (right-hand board) in all other provinces.

	Number of Combinations Having Cluster Number k	
k	7×7 Square	$10 \times 5 - 1$ Rectangle
1	58,620	49,457
2	669,462	636,783
3	2,775,454	2,709,248
4	5,093,678	5,090,231
5	4,166,304	4,237,116
6	1,220,298	1,260,981
Total	13,983,816	13,983,816

Table 7.7 Distribution of the numbers of clusters for the 6/49 lottery game boards of Figures 7.3 and 7.7.

For two shapes of game boards in a 6/49 lottery—namely, a quadratic 7×7 square game board and the rectangular game board shown in Figure 7.6 (for lack of a better term, the latter will be called a 10 × 5 − 1 rectangle)— the number of combinations having cluster number k ($k = 1, 2, \ldots, 6$) is given in Table 7.7.

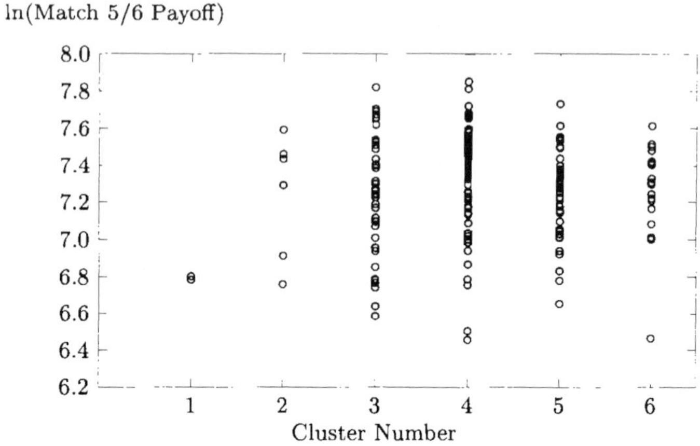

Figure 7.8 Scatterplot of the numbers of clusters versus the ln(match 5/6 payoff) for the Florida State Lottery 6/49, 1993–1996.

Figure 7.8 shows a scatterplot of the cluster numbers (horizontal axis) against the natural logarithms (vertical axis) of the match 5/6 prize payouts. Note that the two combinations having cluster numbers of 1 exhibit fairly low payouts and that the largest match 5/6 payoffs are for combinations having cluster numbers of 3, 4, and 5. Furthermore, the maximum payouts seem to decrease for combinations having cluster numbers of 6. Given all this, you should discard any combination that has an extreme cluster number of 6 or 1. This recommendation eliminates many geometric patterns (including the "Right Angle" in Figure 5.5, the "Y" in Figure 5.6, and the "Horseshoe" in Figure 5.4) that have not yet been discarded by the standing assumption that the AC value should be at least 6. Interestingly, the AC values of the above-mentioned two combinations that have cluster numbers of

1 (that is, 34, 38, 42, 44, 45, 48 and 39, 42, 44, 45, 48, 49) are 5 and 4.

7.5 Played Numbers

Factors such as edge numbers, cluster numbers, arithmetic complexities (ACs), and the sums of numbers of combinations have more or less pronounced influences on payouts. It would be much more interesting, however, to have more direct information on the popularity of combinations among players. Not unexpectedly, the organizers of lotteries don't give out information on player choice, with some notable exceptions.

One of these exceptions is a data set consisting of 16,862,596 combinations chosen for the February 10, 1990, draw of the Swiss Lotto 6/45.[1] One important piece of information that may be gained from such data is how often above or below average each individual number was played. Note that since each of these combinations is made up of six numbers, 101,175,576 ($= 6 \times 16,862,596$) numbers were chosen altogether. Consequently, each of the 45 numbers was selected 2,248,346.13 ($= 101,175,576/45$) times *on average*. For example, the number 21 was chosen 2.909. 924 times, which means that it was selected far more often than average. On the other hand, there were fairly unpopular numbers, such as 44, which was chosen 1,677,861 times (that is, much less often than the average value of 2,248,346.13). Dividing the frequency of each individual number played for this draw of the Swiss Lotto by its average number yields the *relative popularity of each individual number*. Thus, the popularity of 21 is

$$\frac{2,909,924}{2,248,346.13} \approx 1.294.$$

In other words, 21 was selected 29.4% more often than average. The popularity of each number is given in Table 7.8. As you can

[1]These data were generously made available to Hans Riedwyl by the Swiss Lottery Corporation.

see, the most popular number, 9, was chosen 38% more often than average, and the least popular number, 43, was selected 39% less often than average.

Number/ Popularity		Number/ Popularity		Number/ Popularity		Number/ Popularity		Number/ Popularity	
1	0.911	2	0.958	3	1.098	4	1.003	5	1.011
6	1.014	7	0.992	8	1.172	9	1.381	10	1.111
11	1.152	12	0.900	13	0.979	14	1.075	15	1.219
16	1.130	17	1.165	18	0.909	19	0.974	20	1.075
21	1.294	22	1.216	23	1.077	24	0.894	25	0.839
26	1.105	27	1.284	28	1.197	29	1.077	30	0.829
31	0.770	32	1.050	33	1.165	34	1.035	35	0.973
36	0.761	37	0.687	38	0.914	39	0.951	40	0.987
41	0.820	42	0.733	43	0.612	44	0.746	45	0.755

Table 7.8 Popularities of individual numbers for 16,862,596 manually chosen combinations for the February 10, 1990, draw of the Swiss Lotto 6/45.

Figure 7.9 Popularities of individual numbers (indicated by different shades of grey) for 16,862,596 manually chosen combinations for the February, 10, 1990, draw of the Swiss Lotto 6/45.

Figure 7.9 shows the popularity values of Table 7.8 on the Swiss game board. *Popularity* is illustrated by means of different shades of grey. Dark grey corresponds to any number whose popularity is at least 1.07. Medium grey is used for any number with a popularity value between 0.93 and 1.07 (namely, numbers of about average popularity). Each unpopular number, having a popularity of at most 0.93, is indicated using light grey. Observe that, loosely speaking, popularity decreases from the top to the bottom of the game board, and from the middle to the edges.

In addition to the very popular numbers and 9, 21, and 27, there is a "hot rectangle" in the middle of the game board, formed by the numbers 8, 11, 26, and 29. Note that the unpopular numbers are basically those in the first and last columns as well as those in the last row.

The popularities of individual numbers can be used to construct a measure of popularity for a given combination of numbers. Obviously, choosing combinations based on such measures can greatly influence the sizes of any payouts received. You can do the same investigation, provided that your lottery corporation will provide information on players' preferences for individual numbers. If such information on the popularities of individual numbers is available, use this simple method to assess the popularity of a combination using that of the individual numbers: Take the *product* of the popularity of the six individual numbers. For example, the combination 1, 12, 13, 34, 44, 45 can be assigned the popularity

$$0.911 \times 0.900 \times 0.979 \times 1.035 \times 0.746 \times 0.755 \approx 0.468.$$

(An *urn model* for choosing lottery combinations that reflects the popularities of individual numbers is given in Chapter 8, Section 8.15.)

Figure 7.10 illustrates the usefulness of this popularity measure in evaluating match 5/6 payouts. It shows a scatterplot of 52 points, each point indicating one draw of the Swiss Lotto 6/45 in the year 1996. The popularity of each winning combination (horizontal axis) is plotted against its match 5/6 payout, given in

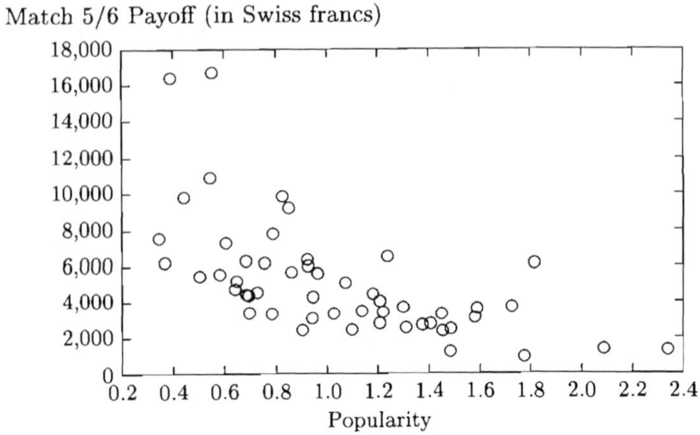

Figure 7.10 Scatterplot of the popularity of each winning combination versus its match 5/6 payoff, Swiss Lotto 6/45, 1996.

Swiss francs (vertical axis). Note that the popularities of the 52 winning combinations ranges from about 0.3 to 2.4 and that, as could have been expected, the match 5/6 payoff tends to decrease with increasing popularity. If a logarithmic transformation is performed for both the popularities and the match 5/6 payoffs, a new scatterplot is obtained, as shown in Figure 7.11. In particular, observe the very strong relationship between the natural logarithm of the popularity measure introduced above and that of the match 5/6 payouts. Indeed, about 53% of the variability of the ln(match 5/6) payoffs can be explained by the measure of popularity. Even though the individual popularities of numbers reflects players' preferences for only one draw during 1990, they still had a great influence on the payoffs of draws that took place even six years later. In sum, while the popularities of numbers varies to some extent over time, it is nevertheless remarkably stable. This has also been observed by Ziemba (1986), who investigated the influence of individual number popularities on payouts for the Canadian Lotto 6/49.

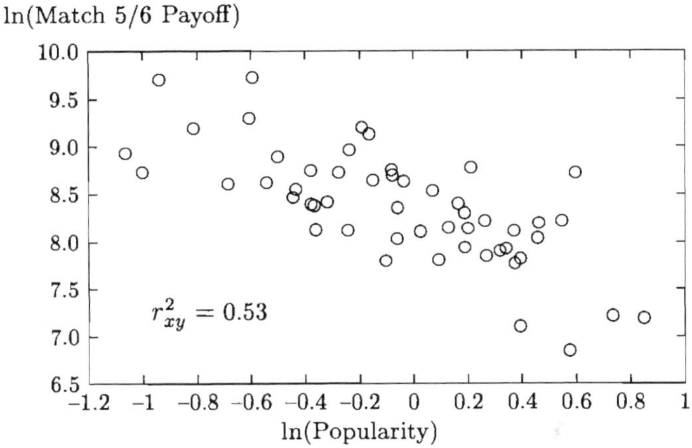

Figure 7.11 Scatterplot of the ln(popularity) of each winning combination versus the ln(match 5/6 payoff) for the Swiss Lotto 6/45, 1996.

7.6 Played Number Combinations

Suppose you have complete information on the number of times that each combination has been selected for some draw—say draw X—of the lottery of your state, and you want to use this information for intelligent play. If your interest is in match 5/6 payoffs, you might choose some combination of numbers and look up how many match 5/6 prize winners this combination would have produced, had it been the winning combination of draw X. For lack of a better term, call this number the *match* 5/6 *popularity of the combination.* For illustration, use the data set of 16,862,596 combinations played for the February 10, 1990 (draw X) draw of the Swiss Lotto 6/45. In that data set, you would find that the combination 2, 12, 19, 30, 40, 42 produced 383 match 5/6 winners in the draw of February 10, 1990. (This was the first winning combination of 1996, by the way.) Thus, the match 5/6 popularity of this combination is 383. Likewise, you would find that the match 5/6 popularity of the combination 4, 8, 14, 21, 22, 32 is 1015, which shows that the latter combination was much more popular, at least for draw X.

At this point, you might ask whether there is some sort of benchmark value which represents the average match 5/6 popularity for draw X. You might also wonder about the predictive value of so-called *snapshot* match 5/6 popularities, which are based on only single draws of lotteries.

As for the first question, remember that in a 6/45 lottery the probability of matching five numbers but not the bonus number is 1 in 35,724 (see Table 3.2). This means that only 1 out of 35,724 randomly chosen combinations will match five of the six winning numbers. Since the number of combinations played for draw X of the Swiss Lotto was $16,862,596$, there would be about $472 \approx 16,862,596/35,724$ match 5/6 winners if all of the combinations were selected at random. Consequently, the number 472 is a benchmark average value for the number of match 5/6 winners, if a total of 16,862,596 combinations is played for a 6/45 lottery. Comparing the match 5/6 popularities of 383 and 1015 with this benchmark value of 472, you can see that the 5/6 popularity value of the first combination was below average, whereas that of the second combination was more than two times above average.

To demonstrate the predictive value of snapshot match 5/6 popularities for match 5/6 payouts in future draws, Figure 7.12 shows a scatterplot of 52 points, each point indicating 1 of the 52 winning combinations for the Swiss Lotto 6/45 in the year 1996. For each combination, the match 5/6 popularity (horizontal axis) is plotted against its match 5/6 payoff, which tends to decrease with increasing match 5/6 popularity. Figure 7.13 shows a plot of the natural logarithm of the match 5/6 popularity against the natural logarithm of the match 5/6 payoffs. Compared with Figure 7.11, the strength of the linear relationship between the logarithm of the match 5/6 popularity values (which are based on one draw of the year 1990) and the ln(match 5/6 payouts) observed *six* years later is striking.

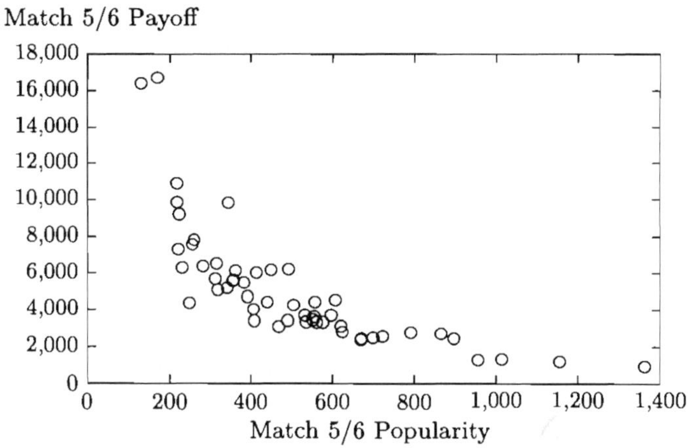

Figure 7.12 Scatterplot of the match 5/6 popularity versus the match 5/6 payoffs for the Swiss Lotto 6/45, 1996.

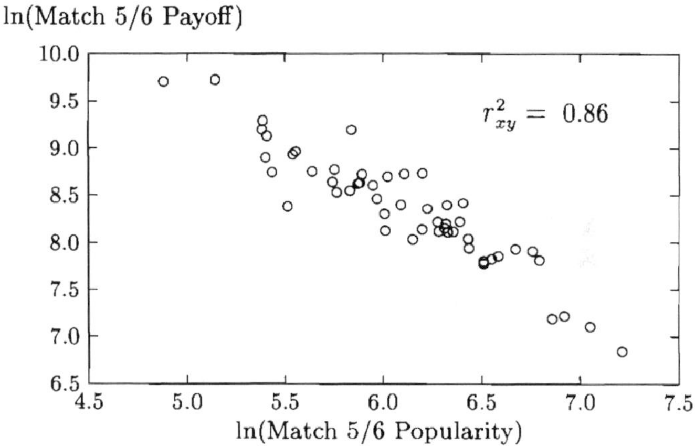

Figure 7.13 Scatterplot of the ln(match 5/6 popularity) versus the ln(match 5/6 payoffs) for the Swiss Lotto 6/45, 1996.

Clearly, anyone who has data on the number of times each combination has been chosen for a draw of the lottery in his or her state may use this information to single out the most underplayed combinations and then choose them for subsequent play. So imagine the interesting situation that would arise if lottery commissions divulged complete information on players' choices, for example, offering such information on either compact disc or the World Wide Web. If this information were so widely available, intelligent play would be much more difficult and perhaps hardly possible.

7.7 A Quick Strategy for Intelligent Play

Assume that your lottery commission doesn't give out information on players' preferences to the public, which is normally the case. That means that you will have to summarize your knowledge of extreme payouts, the various foolish styles of play, and the influences of certain factors on prize payouts to formulate a strategy for intelligent play that is easy to use in practice. Such a strategy is the focus of this section.

For a general r/s lottery, the recommendation is to play only combinations with a large sum of numbers and a high AC value which, at the same time, have a cluster number greater than 1, but less than r, and an edge number of at least $r/2$, but less than r. Since being a "large sum of numbers" depends on the values of r and s, a possible strategy that is applicable for a general number lottery is to generate many (say, 100) combinations purely at random and then, in a first step, to discard 75% of them by considering only those 25 combinations with the largest sums of numbers observed. Each of these combinations is played only if its AC value is at least 6 (say), and if its cluster and edge number satisfy the conditions stated earlier. Of course, more detailed recommendations are possible for specific values of r and s. Understand, however, that the following "recipe" for the special case $r = 6$ and $s = 49$ *is only one of many conceivable intelligent styles of play.*

For a 6/49 lottery, the recommended strategy is the following *restricted random draw*: Choose a combination completely at random (for example, with the help of a deck of cards or a uniform random number generator). Play it, if *all of the following four conditions hold*:

1. The sum of numbers is at least 177.
2. The arithmetic complexity is at least 8.
3. The cluster number is 2, 3, 4, or 5.
4. The edge number is 3, 4, or 5.

If any of these conditions is violated, discard the combination and choose another one. For a 6/49 lottery having a rectangular 10×5 minus 1 game panel, as shown in Figure 7.6 (such as the Florida State Lottery and the British National Lottery), there are exactly 1,521,650 combinations that meet all of the four conditions stated above.

To assess the possible benefits gained from playing this strategy, review yet another experiment: Ten sets of 6/49 lottery tickets were generated, each set consisting of 5,000 combinations chosen according to the restricted random draw strategy stated earlier. These tickets were added fictitiously to those played in the Florida State Lottery and the British National Lottery for the years 1996 and 1995. Their average match 5/6 payoffs are given in Tables 7.9 and 7.10, respectively. Note that playing this

	1996	1995	1996	1995
Florida State Lottery	Match 5/6 Payoff		Match 4/6 Payoff	
Average Payoff	\$1,825	\$1,798	\$73	\$73
Manual Play Average Payoff	\$1,376	\$1,313	\$62	\$59
Performance	32.6%	36.9%	17.7%	23.7%

Table 7.9 Fictitious payoffs of 50,000 randomly chosen combinations (selected using restricted random draw strategy) for the Florida State Lottery 6/49, 1996 and 1995 (in \$U.S.).

U.K. National Lottery	1996 Match 5/6 Payoff	1995 Match 5/6 Payoff	1996 Match 4/6 Payoff	1995 Match 4/6 Payoff
Average Payoff	£2,909	£3,829	£85	£97
Manual Play Average Payoff	£1,575	£1,718	£63	£69
Performance	84.7%	122.9%	34.9%	40.6%

Table 7.10 Fictitious payoffs of 50,000 randomly chosen combinations (selected using restricted random draw strategy) for the British National Lottery 6/49, 1996 and 1995 (in British pounds).

strategy for either the Florida State Lottery or the British National Lottery yields the best performance values observed so far.

7.8 A Regression Strategy for Intelligent Play

If you have a computer and a statistics software package, there is an alternative strategy to the restricted random draw studied in the previous section. It involves using a multiple linear regression model (see Chapter 8, Section 8.11) to exploit the dependence of *observed* payoffs of winning number combinations (published weekly by your lottery corporation) on their main influential factors, such as the AC value, the sum of numbers, and the edge and the cluster number. To this end, consider the data set from Section 7.1 of the 200 match 5/6 payouts of the Florida State Lottery 6/49 during 1993–1996. As a first step, eliminate all combinations having AC values of less than 5 and edge numbers of 1 or 6. With the remaining 177 winning combinations, estimate the predicted value for the dependent variable ln(match 5/6 payoff) on the 4 regressors' AC value, ln(sum of numbers), edge number, and cluster number. A multiple linear regression analysis yields the relationship

$$\begin{aligned} PRED &= 3.4113 + 0.0238 \times AC + 0.7255 \times LS \\ &\quad + 0.0771 \times ED - 0.0403 \times CL, \end{aligned} \tag{7.8.1}$$

where, for short, *PRED* is the predicted value of the dependent variable ln(match 5/6); *AC* is the AC value; *LS* is the natural logarithm of the sum of numbers; *ED* is the edge number; and *CL* denotes the cluster number of the combination.

Figure 7.14 shows a scatterplot of the predicted ln(match 5/6 payouts) (horizontal axis) against the ln(match 5/6 payouts) that were actually observed. In future draws, you may prefer to play number combinations with a high predictive ln(match 5/6 payoff) value.

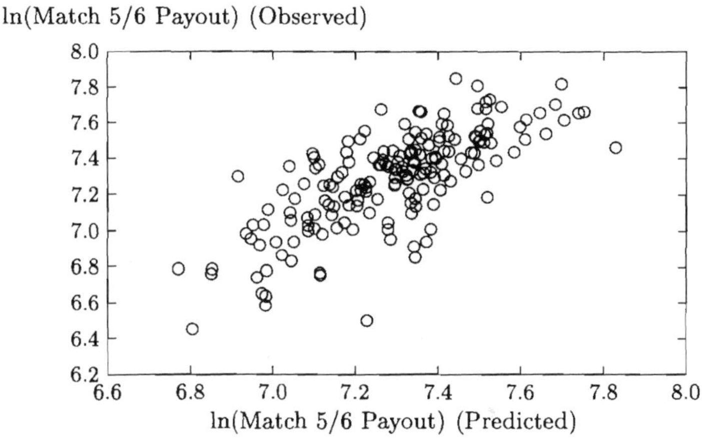

Figure 7.14 Scatterplot of the predicted ln(match 5/6 payouts) against the observed ln(match 5/6 payouts) for the Florida State Lottery 6/49, 1993–1996.

A possible restricted random draw strategy is to choose a combination at random and play it if each of the following 3 conditions holds:

1. The AC value is at least 5.

2. The cluster number is 2, 3, 4, or 5.

3. The predicted ln(match 5/6 payoff), calculated from formula 7.8.1, is at least 7.5.

If you apply this strategy to the 200 winning combinations of the Florida State Lottery during 1993–1996, you will find that exactly 25 of them satisfy each of these 3 conditions. The mean payoff of those 25 combinations is $1,964, and thus somewhat higher than $1,891 (the average payout of those 24 combinations that satisfy the conditions of the quick strategy for intelligent play [see p. 101]).

7.9 Answers to Frequently Asked Questions

This final section deals with 3 frequently raised problems.

1. *Is it profitable to play all of the number combinations to ensure a jackpot win?*

Except for a few extraordinary situations, no. Such a situation is called a *jackpot overlay*, which arises if the first division prize has not been won for several draws, and the players have not yet increased their bettings proportionally to the size of the jackpot. This last point is very important, and can be best understood by thinking of a huge jackpot that is bet on by you, but by no one else! After some carryovers, a jackpot contains millions of dollars contributed by tickets that no longer have any chance of winning. The speculative value of those forfeited dollars, however, decreases with an increasing number of competing tickets.

But even in a situation with a very large jackpot overlay, it is doubtful whether a group of investors playing all the combinations would get even their invested money back (although, in addition to having *some* share of the jackpot, they would win many of the lower division prizes). The key issue is whether a popular or an unpopular number combination is drawn. If unpopular numbers come up, payouts for all prize divisions will be large, and there is a good chance that no one else shares the jackpot with the investors (provided that there are no other players with the same idea for "stealing" the grand prize). In this case, playing all of the combinations could perhaps turn out to be profitable. If a popular combination is drawn, however, payouts are bad, and players run the risk of having to share the jackpot with many others.

The authors agree with Ziemba (1986, p. 57), who gives a detailed analysis of two draws of the Canadian Lotto 6/49 and comes to the conclusion that, at least in general, *it simply does not pay to "buy the pot"*.

2. *Is it better to play the same numbers or skip around?*

The answer to this question is simple: It does not make any difference, since the results of past drawings have no bearing on future winning combinations, at least for the chances of winning one of the various prizes. However, the prospect of winning more, if you do win, should preclude the choice of more or less foolish combinations. Either keep playing your favorite combinations or skip around, but always choose your numbers according to an intelligent style of play.

3. *Does it pay to play a combination two or more times?*

In the long run, as far as the return on each lottery dollar is concerned, the answer is no. To keep things simple, suppose that you play a combination twice, and you hit the jackpot with that combination. If you are the only first division prizewinner, you will win the whole pot regardless of whether you played that combination once, twice, or even more times. In this case, a duplicate ticket is as good as worthless, since it only adds a few cents to the jackpot. If there is one other jackpot winner, you will get a 50% share of the jackpot playing one ticket. If you have a duplicate ticket, the jackpot will have to be split up among three winners, so that your share will be two thirds—about 67% of the jackpot prize. A second ticket only gives you an additional 17% of the jackpot. The result is similar if there are more than two jackpot winners.

Chapter 8

Mathematical Appendix

This appendix contains various mathematical results about number lotteries. The amount of mathematic knowledge will vary with the different topics discussed. All probabilistic results will be derived under the basic assumptions that (1) all winning combinations are equally probable and (2) the results of different (weekly) drawings are statistically independent.

Whenever possible, results will be presented in terms of a general r/s lottery so that you may do your own calculations for any specific problem. Figures or numbers will mostly refer to the case $r = 6$ and $s = 49$.

8.1 Probabilities of Winning

Suppose that you choose a combination of numbers in an r/s lottery. To win one of the various prizes, you will have to match some of the numbers selected in the draw. In addition, some prize structures require the bonus number to be one of your chosen numbers, at least for certain prize levels.

The probability of winning the prize of a specified category may be expressed in terms of binomial coefficients, which involve the *factorials*

$$\begin{aligned} m! &= 1 \cdot 2 \cdot \ldots \cdot (m-1) \cdot m \qquad (m = 1, 2, \ldots), \\ 0! &= 1. \end{aligned}$$

The *binomial coefficient*

$$\binom{n}{k} = \frac{n!}{k! \cdot (n-k)!} \qquad (k = 0, 1, \ldots, n)$$

gives the number of ways of choosing k numbers from among $1, 2, \ldots, n$. For example, there are

$$\binom{4}{2} = \frac{4!}{2! \cdot (4-2)!} = \frac{24}{2 \cdot 2} = 6$$

ways of choosing two numbers out of $1, 2, 3, 4$—namely,

$$1,2 \qquad 1,3 \qquad 1,4 \qquad 2,3 \qquad 2,4 \qquad 3,4.$$

In an r/s lottery, the probability P_k of matching k of the r main numbers is given by the *hypergeometric distribution* (see Section 8.12)

$$P_k = \frac{\binom{r}{k} \cdot \binom{s-r}{r-k}}{\binom{s}{r}} \qquad (k = 0, 1, 2, \ldots, r). \qquad (8.1.1)$$

The probability $P_{k,+}$ of matching k of the r main numbers *plus the bonus number* is

$$P_{k,+} = P_k \cdot \frac{r-k}{s-r} \qquad (k = 0, 1, 2, \ldots, r-1). \qquad (8.1.2)$$

The probability $P_{k,-}$ of matching k of the r main numbers *but not the bonus number* is

$$P_{k,-} = P_k \cdot \frac{s+k-2r}{s-r} \qquad (k = 0, 1, 2, \ldots, r). \qquad (8.1.3)$$

To understand formula (8.1.1), note that the denominator $\binom{s}{r}$ is the number of all possible combinations in an r/s lottery. To match k of r, k of the r numbers you pick must be from among the r main winning numbers, and the other $r-k$ must be from among the $s - r$ nonwinning numbers. There are $\binom{r}{k}$ ways to choose k numbers from r, and for each fixed choice, there are $\binom{s-r}{r-k}$ ways to choose $r - k$ numbers from $s - r$. Thus, P_k may be regarded as the relative frequency of all combinations matching k of your chosen numbers.

Given the r main winning numbers, there are $s - r$ equally likely outcomes for the bonus number. If you know that you have matched k winning numbers (just before observing the drawing of the bonus number!), there are $r - k$ *favorable* and $s - r - (r - k)$ *unfavorable* bonus numbers for you. Thus, the factor to the right of P_k in (8.1.2) stands for the conditional probability of matching the bonus number, given that you match k of the winning numbers. In the same way, the factor to the right of P_k in (8.1.3) stands for the probability of *not* matching the bonus number, given that you match k of the winning numbers. Since matching k numbers always may occur either with or without the bonus number, you have

$$P_k = P_{k,+} + P_{k,-} .$$

The calculation of P_k may be done recursively using the formulae

$$P_0 = \frac{\binom{s-r}{r}}{\binom{s}{r}} = \frac{(s-r)\cdot(s-r-1)\cdot\ \ldots\ \cdot(s-2r+1)}{s\cdot(s-1)\cdot\ \ldots\ \cdot(s-r+1)},$$

$$P_k = P_{k-1}\cdot\frac{(r-k+1)^2}{k\cdot(s-2r+k)} \qquad (k = 1, 2, \ldots, r).$$

Suppose now that you play l different combinations with the property that no two have at least k numbers in common. (This is to ensure that the events of matching k winning numbers on the different game boards will be mutually exclusive.) Then, according to the addition rule for probabilities of mutually exclusive events, the probability of matching k numbers (on at least one game board) is given by P_k figuring in (8.1.1) multiplied by l. In the same way, you have to multiply $P_{k,+}$ figuring in (8.1.2) by the number l of combinations to obtain the probability of matching k winning numbers plus the bonus number; and $P_{k,-}$ figuring in (8.1.3) by l to get the probability of matching k without the bonus number.

Note that the probabilities P_k, $P_{k,+}$, and $P_{k,-}$ refer to the chances of winning prizes of the various categories in only *one* (say, the next) lotto draw. However, your main interest will be in the chances of winning the jackpot at least once within a reasonable period of time. To this end, suppose that you are playing l different combinations in an r/s lottery. Since the probability of *not* winning the jackpot on the next draw is $1 - l \cdot P_r$ and thus for *not winning it on n draws* is $(1 - l \cdot P_r)^n$, the probability for winning the jackpot in an r/s lottery at least once in n draws is given by

$$1 - \left[1 - \frac{l}{\binom{s}{r}}\right]^n . \tag{8.1.4}$$

For example, suppose that you are playing 10 different combinations in the California Super Lotto 6/51, which has two draws per week and thus 2,087 draws in 20 years, if leap years are taken into account. By (8.1.4), the probability of winning the jackpot at least once in 20 years is given by

$$1 - \left[1 - \frac{10}{\binom{51}{6}}\right]^{2{,}087} \approx 0.00158.$$

For further reading on this topic, see Feller (1970), Sections II.6, V.2.

8.2 Frequencies of Winning Numbers

In an r/s lottery, let $\boldsymbol{X_k} = (X_{k,1}, X_{k,2}, \ldots, X_{k,s})$ be the s-dimensional random vector having the components

$$X_{k,j} = \begin{cases} 1, & \text{if number } j \text{ is drawn on draw } k \\ 0, & \text{if number } j \text{ is not drawn on draw } k. \end{cases}$$

Note that

$$H_{n,j} = \sum_{k=1}^{n} X_{k,j} \tag{8.2.1}$$

is the winning frequency of the number j after n draws. Since $X_{1,j}, X_{2,j}, \ldots, X_{n,j}$ are independent and identically distributed (i.i.d.) random variables with expectation $E(X_{k,j}) = r/s$ and variance $Var(X_{k,j}) = r/s \cdot (1 - r/s)$, it follows that the distribution of $H_{n,j}$, for fixed j, is the binomial distribution $Bin(n, r/s)$ (see Section 8.12). That is,

$$P\left(H_{n,j} = l\right) = \binom{n}{l} \cdot \left(\frac{r}{s}\right)^{l} \cdot \left(1 - \frac{r}{s}\right)^{n-l} \qquad (l = 0, \ldots, n).$$

In particular,

$$E\left(H_{n,j}\right) = n \cdot \frac{r}{s}, \qquad Var\left(H_{n,j}\right) = n \cdot \frac{r}{s} \cdot \left(1 - \frac{r}{s}\right).$$

Moreover, the *de Moivre–Laplace Central Limit Theorem* (see, for example, Feller, 1970, p. 186) yields

$$\frac{H_{n,j} - n \cdot \dfrac{r}{s}}{\sqrt{n \cdot \dfrac{r}{s} \cdot \left(1 - \dfrac{r}{s}\right)}} \xrightarrow{\mathcal{D}} \mathcal{N}(0,1) \tag{8.2.2}$$

as $n \to \infty$, where "$\xrightarrow{\mathcal{D}}$" denotes convergence in distribution and $\mathcal{N}(0,1)$ is the standard normal law (see Section 8.12).

From (8.2.2), you may obtain an impression of the probable range of the fluctuation of the winning frequency for any given number j after n draws for large n. Denoting by $z_{\alpha/2}$ the $(1-\alpha/2)$ quantile of the standard normal distribution and letting

$$\sigma_n = \sqrt{n \cdot \frac{r}{s} \cdot \left(1 - \frac{r}{s}\right)}$$

(8.2.2) gives the approximation

$$P\left(n\cdot\frac{r}{s} - z_{\alpha/2}\cdot\sigma_n \leq H_{n,j} \leq n\cdot\frac{r}{s} + z_{\alpha/2}\cdot\sigma_n\right) \approx 1-\alpha \quad (8.2.3)$$

for large n.

It is very important to give a correct interpretation of (8.2.3). To this end, you must fix some number j *in advance* (that is, before observing lotto draws), and then you must consider *n draws of an r/s lottery as one random experiment.* Then, as an outcome of that experiment, the observed winning frequency $h_{n,j}$ of the number j may or may not satisfy the inequalities

$$n \cdot \frac{r}{s} - z_{\alpha/2} \cdot \sigma_n \leq h_{n,j} \leq n \cdot \frac{r}{s} + z_{\alpha/2} \cdot \sigma_n. \quad (8.2.4)$$

The following is an interpretation of (8.2.3) in terms of the *law of large numbers.* If you repeatedly observe the winning frequency of j from n lotto draws, then in the long run (8.2.4) will be satisfied in approximately $(1-\alpha)\cdot 100\%$ of all cases, where a *case* means n lotto draws. As an example, let $\alpha = 0.05$ (in this case, $z_{\alpha/2} = 1.96$), and consider the California Super Lotto 6/51 after 500 draws (that is, $r = 6$, $s = 51$, and $n = 500$). You then have $n \cdot r/s \approx 58.82$, $\sigma_n \approx 7.20$, $n \cdot r/s - z_{\alpha/2} \cdot \sigma_n \approx 44.70$, and $n \cdot r/s + z_{\alpha/2} \cdot \sigma_n \approx 72.94$.

Since $H_{n,j}$ is integer valued, it follows from (8.2.3) that

$$P(45 \leq H_{500,j} \leq 72) \approx 0.95.$$

In the discussion of the correct interpretation of (8.2.3), it was stressed that the number j has to be fixed in advance. The crucial point is that having observed the winning frequencies of all numbers after n draws, people are usually interested in the "hot" or "cold" numbers—namely, those with the largest or smallest winning frequencies, respectively. Since having the largest or smallest winning frequency is determined by all winning frequencies, a probabilistic analysis of the random extreme frequencies

$$Max_n = \max_{1 \leq j \leq s} H_{n,j} \tag{8.2.5}$$

and

$$Min_n = \min_{1 \leq j \leq s} H_{n,j} \tag{8.2.6}$$

requires the study of the joint probabilistic behavior of the random frequencies $H_{n,1}, H_{n,2}, \ldots, H_{n,s}$ defined in (8.2.1). To this end, use the *multivariate central limit theorem* (see, for instance, Feller, 1966, p. 254). According to this theorem, if $\boldsymbol{Y}_1$, $\boldsymbol{Y}_2$, ... is a sequence of i.i.d. s–dimensional random vectors with expectation vector $\boldsymbol{a}$ and covariance matrix Σ, then

$$\frac{1}{\sqrt{n}}\left(\sum_{k=1}^{n} \mathbf{Y}_k - n \cdot \mathbf{a}\right) \xrightarrow{\mathcal{D}} \mathcal{N}_s(\mathbf{0}, \Sigma) \tag{8.2.7}$$

as $n \to \infty$, where $\mathcal{N}_s(\mathbf{0}, \Sigma)$ is a centered s–dimensional normal distribution with covariance matrix Σ (see 8.12).

In what follows,

$$\boldsymbol{Y}_k = (Y_{k,1}, \ldots, Y_{k,s}) = \frac{s}{\sqrt{r \cdot (s-r)}} \cdot \boldsymbol{X}_k,$$

that is,

$$Y_{k,j} = \frac{s}{\sqrt{r(s-r)}} \cdot X_{k,j} \qquad (k \geq 1, 1 \leq j \leq s). \tag{8.2.8}$$

Since $E(X_{k,j}) = r/s$ and $Var(X_{k,j}) = r/s \cdot (1 - r/s)$, it follows that

$$E(Y_{k,j}) = \sqrt{\frac{r}{s-r}}, \qquad Var(Y_{k,j}) = 1.$$

Furthermore, from

$$E(X_{k,i} \cdot X_{k,j}) = P(X_{k,i} = 1, X_{k,j} = 1) = \frac{r}{s} \cdot \frac{r-1}{s-1}$$

for $i \neq j$ you obtain the covariance

$$Cov(X_{k,i}, X_{k,j}) = E(X_{k,i} \cdot X_{k,j}) - \left(\frac{r}{s}\right)^2 = -\frac{r \cdot (s-r)}{s^2 \cdot (s-1)}$$

and thus $Cov(Y_{k,i}, Y_{k,j}) = -1/(s-1)$, if $1 \leq i \neq j \leq s$). In other words, $\boldsymbol{Y}_k$ has expectation vector

$$\boldsymbol{a} = \left(\sqrt{\frac{r}{s-r}}, \sqrt{\frac{r}{s-r}}, \ldots, \sqrt{\frac{r}{s-r}}\right)$$

and covariance matrix

$$\sum = (\sigma_{ij})_{1 \leq i,j \leq s}, \quad \sigma_{ii} = 1, \quad \sigma_{ij} = -\frac{1}{s-1} \ (i \neq j). \quad (8.2.9)$$

Since by (8.2.1) and (8.2.8),

$$\sum_{k=1}^{n} Y_{k,j} = \frac{s}{\sqrt{r \cdot (s-r)}} H_{n,j},$$

(8.2.7) yields

$$\frac{c}{\sqrt{n}} \cdot \left(H_{n,1} - \frac{n \cdot r}{s}, \ \cdots \ , H_{n,s} - \frac{n \cdot r}{s}\right) \xrightarrow{\mathcal{D}} \boldsymbol{T} \qquad (8.2.10)$$

as $n \to \infty$, where $c = s/\sqrt{r \cdot (s-r)}$, and $\boldsymbol{T} = (T_1, \ldots, T_s)$ denotes a random vector with the normal distribution $\mathcal{N}_s(\mathbf{0}, \Sigma)$.

The important message of (8.2.10) is that the random vector

$$\left(H_{n,1} - \frac{n \cdot r}{s}; H_{n,2} - \frac{n \cdot r}{s}; \ldots; H_{n,s} - \frac{n \cdot r}{s}\right)$$

of the fluctuations of the winning frequencies around their common mean value has to be divided by $\sqrt{n}$ in order to achieve a fixed limiting stochastic behavior, as $n \to \infty$. In other words, these fluctuations are of the order $\sqrt{n}$ and thus become larger and larger as n increases.

Note that the limit distribution figuring in (8.2.10) is degenerate. You have

$$H_{n,1} + H_{n,2} + \cdots + H_{n,s} = n \cdot r$$

and thus $P(T_1 + T_2 + \cdots + T_s = 0) = 1$.

From (8.2.10) and the *continuous mapping theorem* (see, for instance, Billingsley, 1968, p. 30), it follows that

$$\frac{1}{\sqrt{n}} \cdot \frac{s}{\sqrt{r(s-r)}} \cdot \left(Max_n - \frac{n \cdot r}{s}\right) \xrightarrow{\mathcal{D}} \max_{1 \le j \le s} T_j,$$

$$\frac{1}{\sqrt{n}} \cdot \frac{s}{\sqrt{r(s-r)}} \cdot \left(Min_n - \frac{n \cdot r}{s}\right) \xrightarrow{\mathcal{D}} \min_{1 \le j \le s} T_j,$$

and

$$\frac{1}{\sqrt{n}} \cdot \frac{s}{\sqrt{r(s-r)}} \cdot (Max_n - Min_n) \xrightarrow{\mathcal{D}} \max_{1 \le j \le s} T_j - \min_{1 \le j \le s} T_j.$$

In particular, the range of the winning frequencies tends to increase with the square root of the number of draws.

8.3 Testing Equiprobability of Lotto Numbers

For a rough check of the assumption that each combination has the same chance of winning, use the test statistic

$$C_n^2 = \frac{s-1}{s-r} \cdot \sum_{j=1}^{s} \frac{\left(H_{n,j} - \frac{n \cdot r}{s}\right)^2}{\frac{n \cdot r}{s}} \qquad (8.3.1)$$

(see Morgenstern, 1979). Note that for each j, C_n^2 compares the winning frequency $H_{n,j}$ of the number j after n draws with its expected value $n \cdot r/s$ under the hypothesis of a uniform distribution of all combinations (see Section 8.2). Under this hypothesis, the limit distribution of C_n^2, as $n \to \infty$, is a chi square distribution with $s-1$ degrees of freedom (see Section 8.12). This is most easily seen by applying the function $g(t_1, \ldots, t_s) = (s-1)/s \cdot \sum_{j=1}^{s} t_j^2$ to the vector on the left-hand side of (8.2.10) (the result is C_n^2) and using the continuous mapping theorem. Since the covariance matrix Σ figuring in (8.2.8) has rank $s - 1$ and satisfies $\Sigma^2 = s/(s-1) \cdot \Sigma$, the limiting C_{s-1}^2 distribution is a consequence of Theorem 4.7.1 of Graybill (1976). Note that the correction factor $(s-1)/(s-r)$ in (8.3.1) accounts for the fact that the set of r consecutively drawn numbers is chosen without replacement.

The chi square test of equiprobability of the lotto numbers rejects the hypothesis of a uniform distribution of all combinations for large values of C_n^2. To see this test in action, consider the observed frequencies of winning numbers for the Canadian Lotto 6/49 after $n = 1,220$ draws, as given in Figure 5.3. For these frequencies, the calculated value of C_n^2 is 59.3. Since this value is smaller than 60.9, which is the 90% quantile of the chi square distribution with 48 degrees of freedom, there are no objections against the fairness of the drawing mechanism.

It must be stressed that the chi square test, as described above, is only able to detect deviations from the assumption that each number has the same probability ($= r/s$) of being drawn. This assumption follows readily from the hypothesis of a uniform distribution over all combinations. However, there are nonuniform probability distributions over the set of all combinations that nevertheless entail the same (marginal) drawing probability for each number.

For chi square tests of the equiprobability of all pairs of lotto numbers and related topics, see Joe (1993). Henze (1982) investigates the behavior of C_n^2 under suitable alternatives to the assumption of a uniform distribution over all combinations.

8.4 First Repetition of a Winning Combination

For a general r/s lottery starting with its first draw, let U be the number of draws until some previously drawn combination shows up again. (There may be even a coincidence of bonus numbers, but this is not required.) Writing

$$c = \binom{s}{r}$$

for the number of all possible combinations in an r/s lottery, regard each combination as one of c *cells*, numbered from 1 to c. A simple numbering is given by the lexicographic order, which for the case $r = 6$, $s = 49$ takes the form

$$\begin{array}{cccccccc}
1 & 2 & 3 & 4 & 5 & 6 & \longrightarrow & 1 \\
1 & 2 & 3 & 4 & 5 & 7 & \longrightarrow & 2 \\
1 & 2 & 3 & 4 & 5 & 8 & \longrightarrow & 3 \\
 & & \vdots & & & & \vdots & \vdots \\
44 & 45 & 46 & 47 & 48 & 49 & \longrightarrow & 13{,}983{,}816 \ (= c)
\end{array} .$$

With this interpretation, each lotto draw may be considered as a placement of a marble into one of c equiprobable cells. The random variable U is then the number of marbles needed until there is a *collision*, in the sense that a marble is placed into a cell that is already occupied. Obviously, U takes one of the possible values $2, 3, \ldots, c, c+1$. You have

$$P(U \geq k+1) = \frac{c \cdot (c-1) \cdot \ \ldots \ \cdot (c-k+1)}{c^k} \qquad (8.4.1)$$

$(k = 1, \ldots, c+1)$ and

$$\begin{aligned}
P(U \leq k) &= 1 - P(U \geq k+1) \\
&= 1 - \prod_{j=1}^{k-1} \left(1 - \frac{j}{c}\right) \qquad (k = 1, \ldots, c+1)
\end{aligned}$$

(see, for instance, Feller, 1970, p. 47).

For the case $r = 6$, $s = 49$, some values of $P(U \leq k)$ are given in Table 8.1.

k	$P(U \leq k)$	k	$P(U \leq k)$
500	0.0089	6,500	0.7792
1,000	0.0351	7,000	0.8266
1,500	0.0773	7,500	0.8662
2,000	0.1332	8,000	0.8986
2,500	0.2002	8,500	0.9245
3,000	0.2751	9,000	0.9448
3,500	0.3546	9,500	0.9603
4,000	0.4356	10,000	0.9720
4,500	0.5152	10,500	0.9806
5,000	0.5909	11,000	0.9868
5,500	0.6609	11,500	0.9912
6,000	0.7240	12,000	0.9942

Table 8.1 Values of $P(U \leq k)$ for the first repetition waiting time U in a 6/49 lottery.

Looking at Table 8.1, you may be astonished at first upon realizing that, in spite of the huge number of different combinations (13,983,816), a favorable bet would be to claim that the first repetition of some previously drawn combination will be after 4.500 draws, at most. This paradoxical phenomenon is nothing but a variant of the classical *birthday problem* (see, for example, Feller, 1970, p. 47). In that context, you have $c = 365$ cells representing the different days of a year, and placing marbles means identifying people's birthdays.

A good approximation to the expected value of U is given by

$$E(U) \approx \sqrt{\frac{\pi}{2} \cdot c} + \frac{2}{3} \qquad (8.4.2)$$

(see, for instance, Blom, Holst, & Sandell, 1994, p. 80). For a 6/49 lottery, this yields $E(U) \approx 4,687$. The message of (8.4.2) is that for a general lotto with c possible combinations, the expected number of draws until the first repetition of some previously shown up set of numbers occurs *is of the order of the square root of* c. Writing U_c instead of U in order to express the dependence of U on the number of different combinations, you have

$$\lim_{c\to\infty} P\left(U_c \leq t \cdot \sqrt{c}\right) = 1 - \exp\left(-\frac{1}{2} \cdot t^2\right), \tag{8.4.3}$$

where $t > 0$. This result, which says that the limit distribution of $c^{-1/2} \cdot U_c$, as $c \to \infty$, is a *Weibull distribution* (see Section 8.12), follows easily from (8.4.1) by putting $k = [t \cdot \sqrt{c}]$ ($[\cdot]$ is the greatest integer function), taking logarithms and using the power expansion of the logarithmic function. Equation (8.4.3) yields the easy-to-use approximation

$$P(U \leq k) \approx 1 - \exp\left(-\frac{k^2}{2 \cdot c}\right). \tag{8.4.4}$$

As an example, consider the New York Lotto 6/54. In this case,

$$c = \binom{54}{6} = \frac{54 \cdot 53 \cdot 52 \cdot 51 \cdot 50 \cdot 49}{1 \cdot 2 \cdot 3 \cdot 4 \cdot 5 \cdot 6} = 25,827,165,$$

and (8.4.4) gives

$$P(U \leq 6,000) \approx 1 - \exp\left(-\frac{6,000^2}{51,654,330}\right) \approx 0.501.$$

Thus, in the New York Lotto, the odds are approximately 2 to 1 of observing the first repetition of some previously drawn combination after 6,000 draws at most.

8.5 Waiting Time to Exhaust Lotto Numbers

A widespread fallacy among players of lotto games is that certain numbers are considered "due" if they have not come up for a long time. To encourage this fallacy, some lottery magazines provide players with current-year winning frequency charts, which start anew at the beginning of each year and show the winning frequencies of each number during that year. An example of a current-year winning frequency chart for a 6/49 Lotto in 1995 after 17 draws is given in Table 8.2. Note that after 17 draws, all but five numbers have already shown up.

1	**2**	**3**	**4**	**5**	**6**	**7**
1	2	1	4	1	2	1
8	**9**	**10**	**11**	**12**	**13**	**14**
3	1	2	2	3	5	–
15	**16**	**17**	**18**	**19**	**20**	**21**
1	–	1	2	1	2	3
22	**23**	**24**	**25**	**26**	**27**	**28**
1	2	2	3	5	2	2
29	**30**	**31**	**32**	**33**	**34**	**35**
–	–	4	4	5	1	1
36	**37**	**38**	**39**	**40**	**41**	**42**
2	4	3	1	1	4	2
43	**44**	**45**	**46**	**47**	**48**	**49**
2	2	3	–	3	2	3

Table 8.2 Typical current-year winning frequency chart after 17 drawings for a 6/49 lottery.

Two obvious questions arise regarding current-year winning frequency charts:

1. How many draws will it take, on average, until each number has come up?

2. What is the probability that each number will have come up at the end of the year (that is, after 52 draws)?

Note that drawing a set of r out of s lotto numbers means forming samples of size r without replacement from a finite set of

s elements which is the classic *collector's problem* situation with group drawings (see, for instance, Stadje, 1990). To answer the questions posed above, introduce the random variable

N = number of *different* winning numbers obtained after k draws,

and the *exhaustion time* random variable

L = number of draws until each number has shown up.

Moreover, let A_j denote the event that j is *not* a winning number in each of the first k draws ($j = 1, \ldots, s$). Since the weekly draws are independent,

$$P(A_{i_1} \cap \ldots \cap A_{i_\nu}) = \left[\frac{\binom{s-\nu}{r}}{\binom{s}{r}}\right]^k$$

$(1 \leq \nu \leq s - r;\ 1 \leq i_1 < \ldots < i_\nu \leq s)$.

Upon noting the crucial equality $\{L \leq k\} = \{N = s\}$, the principle of inclusion and exclusion (see Feller, 1970, p. 99) gives

$$\begin{aligned} P(L \leq k) &= P(N = s) \\ &= 1 - P(N < s) \\ &= 1 - P(A_1 \cup \ldots \cup A_s) \\ &= 1 - \sum_{\nu=1}^{s-r} (-1)^{\nu-1} \cdot \binom{s}{\nu} \cdot \left[\frac{\binom{s-\nu}{r}}{\binom{s}{r}}\right]^k . \end{aligned}$$

Figure 8.1 shows a histogram of the probability distribution of the exhaustion time random variable L for the special case of a 6/49 lottery. Note that the distribution is skewed to the right, which is not surprising, since this is an extreme value problem. (L is a maximum of r waiting times.)

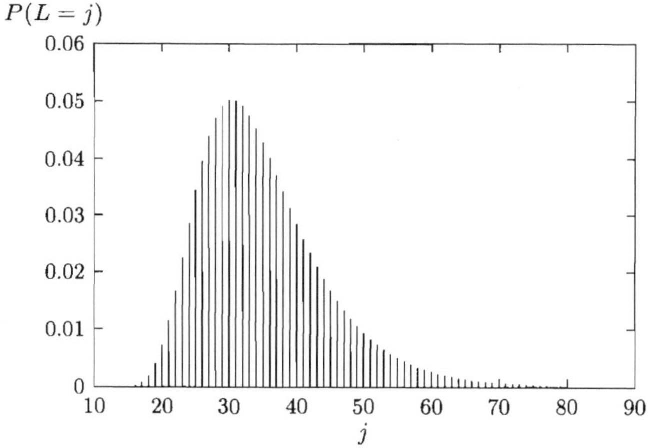

Figure 8.1 Probability distribution of the waiting time to exhaust lotto numbers for a 6/49 lottery.

In particular, you obtain $P(L \leq 52) = 0.946$ and $P(L \leq 53) = 0.953$ for a 6/49 lottery. Thus, for a 6/49 lottery (assuming that the draws are weekly), the probability of exhausting all lotto numbers in the course of a normal year is slightly less than 95%. and it is slightly more than 95% during a leap year.

The expected value of L is given by

$$E(L) = \binom{s}{r} \cdot \sum_{j=0}^{s-1} \frac{(-1)^{s-j-1} \cdot \binom{s}{j}}{\binom{s}{r} - \binom{j}{r}}.$$

In particular, $E(L) \approx 35.08$ if $r = 6, s = 49$. This means that for a 6/49 lottery, each number will have shown up after approximately 35 draws, on average.

For further reading on this topic, consult Leite et. al., 1993.

8.6 Record Waiting Times

Another interesting waiting-time problem is connected with regularly published drawings-since-hit-tables, which indicate how many drawings have taken place since each number was last selected (see Table 5.5). On the occasion of the 2,000th drawing of the German Lotto 6/49, which was celebrated on February 5, 1994, it was mentioned as a curious fact that the number 45 held a "record of outstandingness" in the sense that it had not been drawn between March 3, 1990, and August 31. In other words, it had not shown up for 78 weeks!

To put this result into a probabilistic perspective, you must investigate the distribution of the random variable

$$W_n = \max_{1 \leq j \leq s} W_{n,j},$$

where starting an r/s lottery with its first draw, $W_{n,j}$ is the maximum waiting (outstanding) time (measured in numbers of draws) for the number j in the course of n draws.

To the best of the authors' knowledge, there is no analytic formula for the probability $P(W_n \geq k)$. However, there is an explicit expression for the distribution of the random variable $W_{n,j}$. Neglecting the stochastic dependence of the (exchangeable) random variables $W_{n,1}, \ldots, W_{n,s}$, you can approximate the distribution of W_n by the distribution of the maximum of s independent copies of $W_{n,1}$. This gives

$$\begin{aligned} P(W_n \geq k) &= 1 - P(W_{n,1} < k, W_{n,2} < k, \ldots, W_{n,s} < k) \\ &\approx 1 - P(W_{n,1} < k)^s \qquad (8.6.1) \\ &= 1 - (1 - P(W_{n,1} \geq k))^s . \end{aligned}$$

If you regard the *nonoccurrence* of the number j on the mth draw as a success, the random variable $W_{n,1}$ stands for the maximum number of consecutive successes after n trials (draws) in a sequence of independent trials with probability of success $p =$

$1 - r/s$. An explicit expression for $P(W_{n,1} \geq k)$ is

$$P(W_{n,1} \geq k) \;=\; 1 - \frac{p^{n+1}}{1-p} \cdot F_{n+2}^{(k)}\left(\frac{1-p}{p}\right), \qquad 1 \leq k \leq n,$$

where

$$F_{n+2}^{(k)}(x) \;=\; \sum_{\substack{n_1, n_2, \ldots, n_k \geq 0 \\ n_1 + 2n_2 + \cdots + kn_k = n+1}} \frac{(n_1 + \cdots + n_k)!}{n_1! n_2! \ldots n_k!} \cdot x^{n_1 + \cdots + n_k}$$

denotes the *Fibonacci-type polynomial of order k* (see Philippou & Makri, 1985).

For a numerical calculation of $P(W_{n,1} \geq k)$, you may use the formula

$$P(W_{n,1} \geq k) \;=\; \sum_{m=k}^{n} p_m,$$

where p_m is the probability of the event that on the mth draw, k consecutive successes will be observed for the first time.

Obviously, the sequence $(p_m)_{m \geq 1}$ satisfies the boundary conditions

$$p_1 = \ldots = p_{k-1} = 0, \quad p_k = p^k \tag{8.6.2}$$

and the recursion formula

$$p_j \;=\; q \cdot p_{j-1} + p \cdot q \cdot p_{j-2} + \cdots + p^{k-1} \cdot q \cdot p_{j-k} \tag{8.6.3}$$

$(j \geq k+1)$, where, for short, $q = 1 - p$.

For the case $n = 2,000$, $k = 78$, $r = 6$, and $s = 49$ (that is, $p = 43/49$), (8.6.3) yields

$$P(W_{2000,1} \geq 78) = 0.0088515 \ \ldots,$$

and plugging this result into (8.6.2), you obtain

$$P(W_{2000} \geq 78) \approx 0.35316. \tag{8.6.4}$$

To check the accuracy of this approximation, the situation of 2,000 draws of a 6/49 lottery was simulated 100,000 times, and in each case, the record outstanding time was determined. Figure 8.2 shows a frequency histogram of the observed 100,000 simulated maximum waiting times.

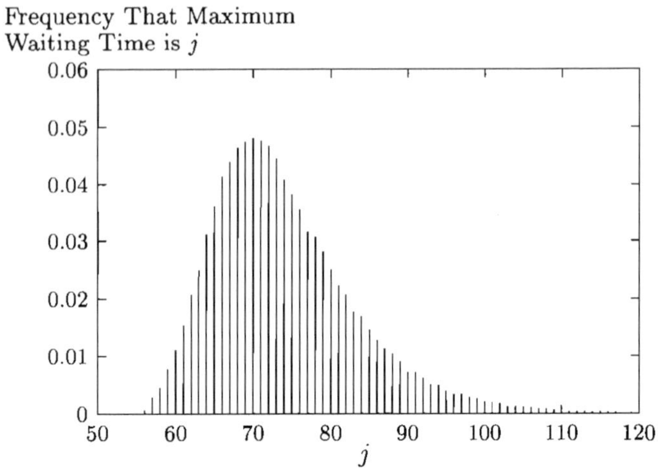

Figure 8.2 Frequency histogram of 100,000 record waiting times for a 6/49 lottery after 2,000 draws.

Apart from the fact that this empirical distribution is skewed (which is typical for a stochastic extreme value problem), it is clear that observing a maximum waiting time of at least 78 weeks *in the course of 2,000 draws* is not at all unusual. In fact, the mean simulated maximum waiting time was 75.5 weeks, and an estimated value of $P(W_{2000} \geq 78)$ based on the observed frequencies is

$$P(W_{2000} \geq 78) \approx 0.351,$$

together with the confidence interval $[0.348,\ 0.354]$ at level 0.95. Obviously, this is in perfect agreement with the theoretical approximation (8.6.4).

For further reading on this subject, see Henze (1998).

8.7 The Distribution of Spaces on Lottery Tickets

In this section, let $T = \{t = (t_1, t_2, \ldots, t_r) : 1 \le t_1 < t_2 < \ldots < t_r \le s\}$ denote the set of all combinations in an r/s lottery. For a combination t, the jth *spacing* is defined as

$$V_j(t) = t_{j+1} - t_j, \qquad j = 1, \ldots, r-1.$$

By means of a certain *shrinking procedure*, the distribution of a *single-spacing* V_j, as well as the distributions of the *smallest spacing*

$$S(t) = \min\{V_j(t) \ : \ j = 1, \ldots, r-1\}$$

and the *largest spacing*

$$L(t) = \max\{V_j(t) \ : \ j = 1, \ldots, r-1\},$$

were derived by Henze (1995). The distribution of a single-spacing is given by

$$P(V_j = k) \ = \ \binom{s-k}{r-1} \Big/ \binom{s}{r}, \qquad 1 \le k \le s+1-r,$$

from which it follows in particular that $E(V_j) = 50/7 \approx 7.4$ and $Var(V_j) = 3225/98 \approx 32.9$ for a 6/49 lottery.

The distribution of the smallest spacing is

$$P(S = k) \ = \ \frac{\binom{s-(r-1)(k-1)}{r}^{+} - \binom{s-(r-1)k}{r}^{+}}{\binom{s}{r}}$$

$(1 \le k \le (s-1)/(r-1))$, where $\binom{m}{n}^{+} = 0$ if $m < 0$ and $\binom{m}{n}^{+} = \binom{m}{n}$, otherwise.

In particular, $P(S = 1) \approx 0.495$ for a 6/49 lottery. Thus, contrary to intuition, the chance of observing at least one pair of consecutive winning numbers (a *twin*) is nearly 50%.

The distribution of the largest spacing L may be obtained by the method of inclusion and exclusion (see Henze, 1995). For $1 \leq k \leq s - r + 1$, you have

$$P(L = k) = \binom{s}{r}^{-1} \sum_{\nu=1}^{r-1} (-1)^{\nu-1} \binom{r-1}{\nu} \cdot \left[\binom{s - \nu(k-1)}{r}^{+} - \binom{s - \nu k}{r}^{+} \right].$$

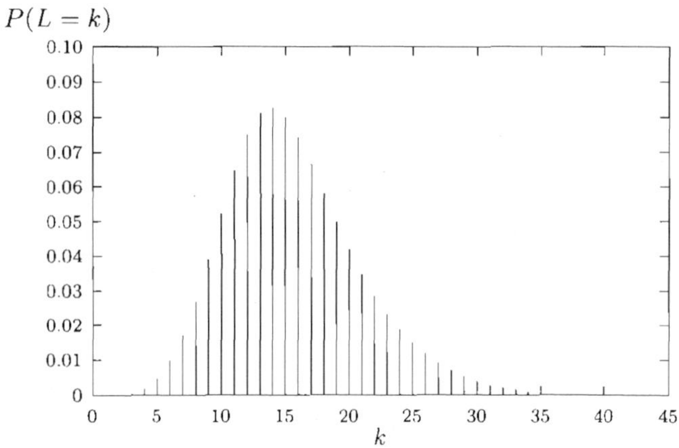

Figure 8.3 Distribution of the largest spacing for a 6/49 lottery.

Figure 8.3 is a frequency chart of the probability distribution of the maximum spacing L for the case of a 6/49 lottery. Note that the distribution is skewed to the right. The mode is 14, whereas the mean largest space is approximately $E(L) \approx 15.64$. Also observe that, again contrary to intuition, a maximum spacing of at least 20 is not at all unusual in a 6/49 lottery.

8.8 The Sum of Winning Numbers

As in Section 8.7, let $T = \{t = (t_1, t_2, \ldots, t_r) : 1 \leq t_1 < t_2 < \ldots < t_r \leq s\}$ be the set of all combinations in an r/s lottery. The distribution of the sum

$$SUM(t) = \sum_{j=1}^{r} t_j$$

of winning numbers is the distribution of the classical two-sample Wilcoxon rank sum statistic with sample sizes r and $s - r$, under the hypothesis of equality of two continuous distribution functions (see Gibbons & Chakraborti, 1992, Section 9.2). The range for SUM is

$$\frac{r \cdot (r+1)}{2} \leq SUM \leq r \cdot s - \frac{r \cdot (r-1)}{2},$$

and

$$E(SUM) = \frac{r \cdot (s+1)}{2}, \quad V(SUM) = \frac{(s-r) \cdot r \cdot (s+1)}{12}.$$

Moreover, the distribution of SUM is symmetric. For the special case of a 6/49 lottery, a frequency histogram of the distribution of SUM is given in Figure 8.4.

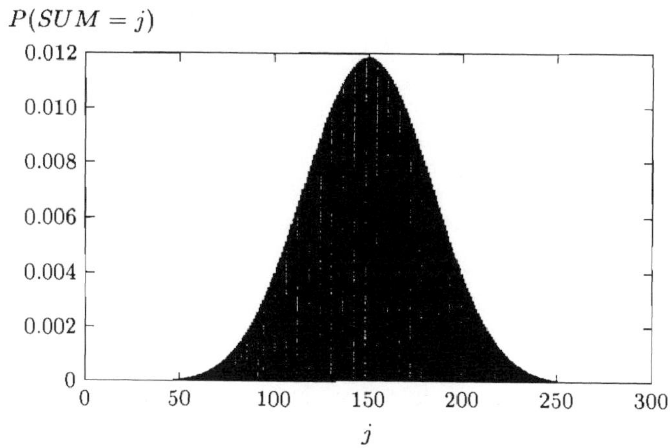

Figure 8.4 Distribution of the sum of winning numbers for a 6/49 lottery.

In this case, $E(SUM) = 150$ and the standard deviation is $\sigma(SUM) = V(SUM)^{1/2} \approx 32.8$. For some selected values of k, Table 8.3 gives the probability $P(SUM \leq k)$ and the number $\#\{SUM \leq k\}$ of combinations t such that $SUM(t) \leq k$.

k	$P(SUM \leq k)$	$\#\{SUM \leq k\}$
55	0.0011	15,121
60	0.0021	28,767
68	0.0051	70.927
75	0.0101	141,179
83	0.0202	282,048
95	0.0486	680.091
107	0.1002	1,400,592
127	0.2511	3,511,072
149	0.4941	6,909,022
150	0.5059	7,074,794
172	0.7489	10,472,744
192	0.8998	12,583,224
204	0.9514	13,303,725
216	0.9798	13,701,768
224	0.9899	13,842,637
231	0.9949	13,912,889
239	0.9979	13,955,049
244	0.9989	13,968,695

Table 8.3 Selected values of $P(SUM \leq k)$ and corresponding numbers of combinations for a 6/49 lottery.

8.9 A Measure of Arithmetic Complexity

As in Section 8.7, let $T = \{t = (t_1, t_2, \ldots, t_r) : 1 \leq t_1 < t_2 < \ldots < t_r \leq s\}$ be the set of all combinations in an r/s lottery. To define a measure $AC(t)$ of arithmetic complexity for the combination t, let $D(t)$ denote the number of all *different* positive differences between any two of the numbers $t_1, \ldots, t_r$. That is, $D(t)$ is the cardinality of the set $\{t_k - t_l : 1 \leq l < k \leq r\}$. For example, if $r = 4$ and $t = (3, 7, 11, 22)$, there are six differences: 4 $(= 7 - 3)$, 4 $(= 11 - 7)$, 11 $(= 22 - 11)$, 8 $(= 11 - 3)$, 15 $(= 22 - 7)$, and 19 $(= 22 - 3)$. Of these, only five are different, so $D(t) = 5$. For the arithmetic progression $t = (12, 14, 16, 18)$, there are only three different differences (namely 2, 4, and 6) so that $D(t) = 3$.

Generally, you have the inequalities $r - 1 \leq D(t) \leq \binom{r}{2}$, with $D(t) = r - 1$ if, and only if, the combination t is an arithmetic progression and thus arithmetically extremely simple. The *measure of arithmetic complexity of* t, denoted by $AC(t)$, is defined by

$$AC(t) = D(t) - r + 1. \tag{8.9.1}$$

Thus, $0 \leq AC(t) \leq \binom{r}{2} - r + 1$, and $AC(t) = 0$ if, and only if, the combination t is an arithmetic progression. Assuming a uniform distribution over all combinations, the probability distribution of the random variable AC is given in Table 8.4 for the case $r = 6, s = 49$.

k	$\#AC = k$	$P(AC = k)$	$P(AC \leq k)$
0	216	0.00002	0.00002
1	1027	0.00007	0.00009
2	10147	0.00073	0.00081
3	34495	0.00247	0.00328
4	141271	0.01010	0.01338
5	335878	0.02402	0.03740
6	1219724	0.08722	0.12463
7	1749352	0.12510	0.24973
8	4542028	0.32481	0.57453
9	2871100	0.20532	0.77985
10	3078578	0.22015	1.00000

Table 8.4 Distribution of the arithmetic complexity variable AC for a 6/49 lottery.

In Table 8.4, the entry $\#AC = k$ is the number of combinations t such that $AC(t) = k$. Note that in a 6/49 lottery, more than 96% of all combinations have an arithmetic complexity of at least six.

8.10 The Correlation Coefficient

A classical measure of association between two variables, X and Y, is the *(Pearson) correlation coefficient*:

$$r_{xy} = \frac{\sum_{i=1}^{n}(x_i - \bar{x}) \cdot (y_i - \bar{y})}{\sqrt{\sum_{i=1}^{n}(x_i - \bar{x})^2 \cdot \sum_{i=1}^{n}(y_i - \bar{y})^2}}.$$

Here, $\bar{x} = n^{-1}\sum_{i=1}^{n} x_i$ and $\bar{y} = n^{-1}\sum_{i=1}^{n} y_i$ are the arithmetic means of $x_1, \ldots, x_n$ and $y_1, \ldots, y_n$, respectively. The values x_i and y_i may be interpreted as the results of the ith measurement on the variables X and Y. In Figures 4.10 and 4.11, the variable X is the sum of the drawn numbers, and Y stands for the match 4 and for the match 5 payoffs, respectively. The term *ith measurement* refers to the ith draw in 1994. Since the minimum of the sum of squared deviations $\sum_{i=1}^{n}(y_i - a - bx_i)^2$, taken over all values of a and b, is given by

$$(1 - r_{xy}^2) \cdot \sum_{i=1}^{n}(y_i - \bar{y})^2,$$

the square r_{xy}^2 of the correlation coefficient (the so-called *coefficient of determination*) may be interpreted as the proportion of the variance of one variable that can be explained by a straight-line dependence on the other variable.

You have $-1 \leq r_{xy} \leq +1$. Furthermore, $r_{xy} = +1$ or $r_{xy} = -1$ if, and only if, the points (x_i, y_i) $(i = 1, \ldots, n)$ lie on a straight line with a positive or a negative slope, respectively.

8.11 Multiple Linear Regression

The model of *multiple linear regression* assumes that the mean μ of a *response* (*dependent*) random variable Y depends linearly on certain *regressors* (*independent variables*) $x_1, \ldots, x_p$. Namely,

$$\mu = \beta_0 + \beta_1 \cdot x_1 + \beta_2 \cdot x_2 + \ldots + \beta_p \cdot x_p,$$

where $\beta_0, \beta_1, \ldots, \beta_p$ are unknown parameters. To estimate $\beta_0, \beta_1, \ldots, \beta_p$, the available data are of the form

$$y_j, x_{1j}, x_{2j}, \ldots, x_{pj} \quad (j = 1, \ldots, n; n \geq p+1), \qquad (8.11.1)$$

where y_j stands for the jth measurement of Y, taken at the regressor values $x_{1j}, \ldots, x_{pj}$. As a function of $\beta_0, \beta_1, \ldots, \beta_p$, the *sum of squares*

$$\sum_{j=1}^{n} (y_j - \beta_0 - \beta_1 \cdot x_{1j} - \ldots - \beta_p \cdot x_{pj})^2$$

assumes its minimum value for the so-called *least squares estimators* $\widehat{\beta_0}, \widehat{\beta_1}, \ldots, \widehat{\beta_p}$ of $\beta_0, \beta_1, \ldots, \beta_p$. Denoting the *predicted* (*fitted*) *values* by $\widehat{y_j} = \widehat{\beta_0} + \widehat{\beta_1} \cdot x_{1j} + \ldots + \widehat{\beta_p} \cdot x_{pj}$, the correlation coefficient (see Section 8.10) of the pairs $(y_j, \widehat{y_j})$ $(j = 1, \ldots, n)$ is called the (empirical) *multiple correlation coefficient* between Y and $x_1, \ldots, x_p$, given the data (8.11.1), and it is denoted by R.

Since

$$R^2 = 1 - \frac{\sum_{j=1}^{n}(y_j - \widehat{y_j})^2}{\sum_{j=1}^{n}(y_j - \bar{y})^2},$$

where $\bar{y}$ is the mean value of $y_1, \ldots, y_n$, the square of R (the so-called *coefficient of determination*) may be interpreted as the proportion of the variability of $y_1, \ldots, y_n$ that is explained by a linear regression on $x_1, \ldots, x_p$. You have $0 \leq R^2 \leq 1$, with R^2 being close to 0 if no linear dependency exists. On the other hand, R^2 is near 1 in case of a strong linear dependency.

8.12 Some Standard Distributions

The Binomial Distribution

Consider a trial with two possible outcomes, *success* and *failure*. The probability of success is denoted by p. If n such trials are performed independently of each other, the total number X of successes has the binomial distribution

$$P(X = k) \;=\; \binom{n}{k} \cdot p^k \cdot (1-p)^{n-k}, \qquad (8.12.1)$$

where $k = 0, 1, \ldots, n$. In short, $X \sim Bin(n, p)$.

Expectation and variance of X are given by

$$E(X) = n \cdot p, \quad Var(X) = n \cdot p \cdot (1-p).$$

The Poisson Distribution

A random variable X is said to have a *Poisson distribution with parameter* λ $(\lambda > 0)$ if

$$P(X = k) \;=\; e^{-\lambda} \cdot \frac{\lambda^k}{k!} \qquad (k = 0, 1, 2, \ldots), \qquad (8.12.2)$$

which is denoted by $X \sim Po(\lambda)$.

Expectation and variance of X are given by

$$E(X) \;=\; Var(X) \;=\; \lambda.$$

The Poisson distribution $Po(\lambda)$ is a good approximation to the binomial distribution $Bin(n, p)$ for large n and small p if $\lambda = n \cdot p$. Namely, you have

$$\binom{n}{k} \cdot p^k \cdot (1-p)^{n-k} \;\approx\; e^{-\lambda} \cdot \frac{\lambda^k}{k!}, \qquad \lambda = n \cdot p \qquad (8.12.3)$$

for large n and small p.

The Hypergeometric Distribution

In a population of s elements, r are *red* and $s-r$ are *black*. If a group of n elements is sampled at random—that is, each of the $\binom{s}{n}$ subsets with n elements has the same chance of being selected—the random number X of red elements in the sample has the *hypergeometric distribution*

$$P(X=k) = \frac{\begin{pmatrix} r \\ k \end{pmatrix} \cdot \begin{pmatrix} s-r \\ n-k \end{pmatrix}}{\begin{pmatrix} s \\ n \end{pmatrix}} \qquad (k=0,1,\ldots,n). \tag{8.12.4}$$

In short, $X \sim Hyp(n,r,s)$.

The Normal Distribution

A continuous random variable X is said to have a *normal distribution with parameters μ and σ^2* ($\sigma^2 > 0$) if its probability density function is given by

$$f(x) \;=\; \frac{1}{\sigma \cdot \sqrt{2\pi}} \cdot \exp\left(-\frac{(x-\mu)^2}{2\sigma^2}\right), \qquad -\infty < x < \infty.$$

In short, $X \sim \mathcal{N}(\mu, \sigma^2)$. In the *degenerate case* (that is, if $P[X = \mu] = 1$ for some μ), you will have a normal distribution with variance 0.

Expectation and variance of X are given by

$$E(X) = \mu, \; Var(X) = \sigma^2.$$

If $\mu = 0$ and $\sigma^2 = 1$, the distribution of X is said to be *standard normal.* The standard normal distribution is the limit distribution of the standardized binomial distribution $Bin(n,p)$ as $n \to \infty$ and of the standardized Poisson distribution $Po(\lambda)$ as $\lambda \to \infty$. Likewise, the standardized hypergeometric distribution $Hyp(n,r,s)$ converges to the standard normal law, provided that n, r, and s tend to infinity in such a way that the variance of the distribution $Hyp(n,r,s)$ tends to infinity as well.

The Multivariate Normal Distribution

An s-dimensional random vector $\boldsymbol{X} = (X_1, \ldots, X_s)$ is said to have an *s-dimensional normal distribution with expectation (vector)* $\boldsymbol{\mu} = (\mu_1, \ldots, \mu_s)$ *and covariance matrix* $\Sigma = (\sigma_{i,j})_{1 \le i,i \le s}$, if for any choice of real numbers $c_1, \ldots, c_s$ the random variable $c_1 X_1 + \ldots + c_s X_s$ has the univariate normal distribution

$$\mathcal{N}\left(\sum_{i=1}^{s} c_i \mu_i, \sum_{i=1}^{s}\sum_{j=1}^{s} c_i c_j \sigma_{i,j}\right).$$

The distribution of $\boldsymbol{X}$ is uniquely determined by $\boldsymbol{\mu}$ and Σ, which is denoted by

$$\boldsymbol{X} \sim \mathcal{N}_s(\boldsymbol{\mu}, \Sigma).$$

If Σ is nonsingular, the distribution of $\boldsymbol{X}$ is called *nondegenerate*; otherwise, it is *degenerate*. A nondegenerate normal distribution has the density

$$f(\boldsymbol{x}) = (2\pi)^{-s/2} \cdot |\Sigma|^{-1/2} \cdot exp\left(-\frac{1}{2}(\boldsymbol{x}-\boldsymbol{\mu})\cdot\Sigma^{-1}\cdot(\boldsymbol{x}-\boldsymbol{\mu})'\right),$$

where $\boldsymbol{x} = (x_1, \ldots, x_s)$ and the prime denotes transpose.

The Chi Square Distribution

Let $X_1, \ldots, X_k$ be independent standard normal random variables. The distribution of the random variable

$$X = X_1^2 + X_2^2 + \ldots + X_k^2$$

is called the *chi square distribution with k degrees of freedom*; in short, $X \sim \chi_k^2$. The density of X is given by

$$f(x) = 2^{-s/2} \cdot \Gamma(s/2)^{-1} \cdot e^{-x/2} \cdot x^{s/2-1}, \qquad x > 0,$$

and $f(x) = 0$ otherwise. Here $\Gamma(u) = \int_0^\infty t^{u-1} e^{-t} dt$ denotes the *Gamma function.*

Expectation and variance of X are given by

$$E(X) = k, \qquad Var(X) = 2 \cdot k.$$

The Weibull Distribution

A continuous random variable X is said to have a *Weibull distribution with parameters* λ and α $(\lambda, \alpha > 0)$ if the density of X is given by

$$f(x) \;=\; \alpha \cdot \lambda \cdot x^{\alpha-1} \cdot e^{-\lambda \cdot x^{\alpha}}, \qquad x > 0,$$

and $f(x) = 0$ otherwise. That is, $X \sim W(\lambda, \alpha)$.

8.13 Mathematical Aspects of Quick-Pick Play

As before, let $T = \{t = (t_1, \ldots, t_r) : 1 \leq t_1 < \ldots < t_r \leq s\}$ denote the set of all possible combinations in an r/s lottery, and let

$$c \;=\; \binom{s}{r}$$

be the number of all such combinations. The *play* on the combination t in a fixed draw is the number of all tickets purchased on the r–tuple t. In what follows, assume that N tickets have been sold for a draw. The mean number of tickets per combination is then given by

$$\lambda \;=\; N/c.$$

Under the assumption that all bets are made independently of each other and completely at random (that is, this is an ideal situation without manual play and a statistically perfect quick-pick machine), the number N_t of bets on a fixed combination t

has the binomial distribution $Bin(N, 1/c)$ (see p. 132). Since N and c are large, the distribution of N_t is approximately Poisson with parameter λ so that

$$P(N_t = k) \approx e^{-\lambda} \cdot \frac{\lambda^k}{k!} \qquad k = 0, 1, 2, \ldots$$

It follows that under the standing assumptions of independence of the drawings and a uniform distribution of winning numbers, the number of jackpot winners would have a Poisson distribution with expectation λ. In particular, the probability of a jackpot rollover would be approximately $P(\text{rollover}) \approx e^{-\lambda}$.

The following argument shows that within a model of independent and identically distributed bets, complete random play is least likely to produce a jackpot rollover. To this end, assume that bets are made independently of each other, but in generalization of the assumption of a uniform distribution over all combinations, combination t is chosen with probability p_t for each single bet, where $p_t \geq 0$ and $\sum_{t \in T} p_t = 1$. By first considering the probability of a jackpot rollover conditional on a fixed winning combination t, you obtain

$$P(\text{jackpot rollover}) = \frac{1}{c} \cdot \sum_{t \in T} (1 - p_t)^N.$$

Using Hölder's inequality (see, for example, Lee, 1972, p. 12)

$$\sum_j |u_j v_j| \leq \left(\sum_j |u_j|^\alpha \right)^{1/\alpha} \cdot \left(\sum_j |v_j|^\beta \right)^{1/\beta}$$

$(1 < \alpha < \infty,\ 1/\alpha + 1/\beta = 1)$ with $j = t,\ u_t = 1 - p_t,\ v_t = 1$ and $\alpha = N$, you obtain

$$\left(1 - \frac{1}{c}\right)^N \leq \frac{1}{c} \cdot \sum_{t \in T} (1 - p_t)^N.$$

This shows that within a model of identically distributed player preferences, complete random play is least likely to produce a jackpot rollover, a result that is intuitively clear.

Let R_k denote the proportion of combinations played k times; namely,

$$R_k = \frac{1}{c} \cdot \sum_{t \in T} 1\{\text{combination } t \text{ is played } k \text{ times}\},$$

where $1\{A\}$ is the indicator function of an event A, which is 1 if A occurs and 0 otherwise.

Under complete random play, R_k has expectation

$$E(R_k) = \binom{N}{k} \cdot \left(\frac{1}{c}\right)^k \cdot \left(1 - \frac{1}{c}\right)^{N-k},$$

which for large N and small k is approximately $E(R_k) \approx e^{-\lambda} \cdot \lambda^k/k!$, where $\lambda = N/c$.

8.14 The Natural Logarithm

The *natural logarithm function*, defined by

$$\ln(x) = \int_1^x \frac{1}{t}\, dt$$

for $x > 0$, is a function that transforms a positive number x to a number $\ln(x)$, as illustrated in Figure 8.5.

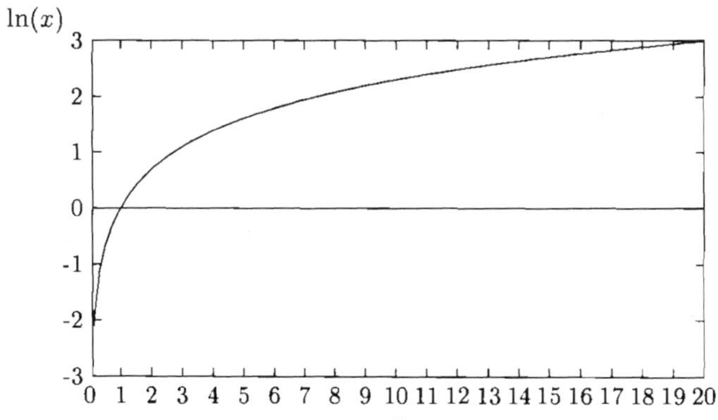

Figure 8.5 The natural logarithm function.

You may obtain the value $\ln(x)$ from your pocket calculator. For example, $\ln(10) = 2.3025\ldots$ and $\ln(1{,}000) = 6.9077\ldots$. The transformation is strictly increasing, and its main properties are

$$\ln(x \cdot y) \;=\; \ln(x) \;+\; \ln(y)$$

and $\ln(e) \;=\; 1$, where $e = \sum_{n=0}^{\infty} 1/n! \approx 2.71828$.

8.15 An Urn Model Reflecting Number Popularity

The following special case of a generalized Pólya-Eggenberger urn model (see, for instance, Johnson & Kotz, 1977, p. 176+ff.) is appropriate for constructing nonuniform distributions over all combinations in an r/s lottery.

Suppose that an urn contains k balls, of which k_j are labeled with the number j, where $j = 1, 2, \ldots, s$. Draw a ball at random, observe its number (say, j_1), and remove all balls showing the number j_1. Then draw a ball at random from the remaining $k - k_{j_1}$ balls, note its number (say, j_2), and remove all balls labeled j_2. Continuing this way until r balls are drawn, you obtain r distinct numbers $j_1, \ldots, j_r$ from among the numbers $1, 2, \ldots, s$. The probability of observing the sequence of numbers $j_1, j_2, \ldots, j_r$ (in this order of succession) is given by

$$\begin{aligned} p(j_1, \ldots, j_r) &= \frac{k_{j_1}}{k} \cdot \frac{k_{j_2}}{k - k_{j_1}} \cdot \ldots \cdot \frac{k_{j_r}}{k - k_{j_1} - k_{j_2} - \ldots - k_{j_{r-1}}} \\ &= \prod_{\nu=1}^{r} p_{j_\nu} \cdot \prod_{\nu=1}^{r-1} \left(1 - p_{j_1} \;-\; \ldots \;-\; p_{j_\nu}\right)^{-1}, \end{aligned}$$

where $p_j \;=\; k_j/k$ is the probability that the first ball drawn has the number j ($j = 1, \ldots, s$).

The probability of obtaining the combination $\{j_1, \ldots, j_r\}$ (in an arbitrary order of succession) is

$$P(\{j_1, \ldots, j_r\}) = \prod_{\nu=1}^{r} p_{j_\nu} \cdot \sum_{r!} \prod_{\nu=1}^{r-1} \left(1 - \sum_{\rho=1}^{\nu} p_{j_\rho}\right)^{-1}, \quad (8.15.1)$$

where the symbol $\sum_{r!}$ denotes summation over all $r!$ permutations of $(j_1, \ldots, j_r)$.

Equation (8.15.1) defines a probability distribution over all $\binom{s}{r}$ combinations in an r/s lottery; that is,

$$\sum_{1 \le j_1 < \ldots < j_r \le s} P(\{j_1, \ldots, j_r\}) = 1.$$

If $p_j = 1/s$ for $j = 1, \ldots, s$, (8.15.1) reduces to

$$P(\{j_1, \ldots, j_r\}) = \left(\frac{1}{s}\right)^r \cdot r! \cdot \prod_{\nu=1}^{r-1} \left(1 - \frac{\nu}{s}\right)^{-1} = \frac{1}{\binom{s}{r}},$$

which is the uniform distribution over all combinations. Writing

$$q_j = s \cdot p_j = \frac{p_j}{1/s}$$

for the probability of obtaining number j in the first draw *relative to the uniform probability* $1/s$, (8.15.1) takes the form

$$P(\{j_1, \ldots, j_r\}) = \frac{\prod_{\nu=1}^{r} q_{j_\nu} \cdot \frac{1}{r!} \cdot \sum_{r!} \left[\frac{\prod_{\rho=1}^{r-1} (s - \rho)}{\prod_{\nu=1}^{r-1} (s - \sum_{\mu=1}^{\nu} q_{j_\mu})} \right]}{\binom{s}{r}}.$$

Note that $q_1, \ldots, q_s$ correspond to the individual number popularities introduced in Section 7.5, and that the deviation of $P(\{j_1, \ldots, j_r\})$ from the uniform distribution $1/\binom{s}{r}$ is mainly due to the product $\prod_{\nu=1}^{r} q_{j_\nu}$.

8.16 A Lotto Random Number Generator

A simple way to select a random (quick-pick) combination in an r/s lottery is to use a *(pseudo) uniform random number generator.* Such number generators are of paramount importance for the simulation of random processes, and even your pocket calculator may have a key labeled *RND* or *RD* which, when pressed, will yield a random number x in the unit interval.

A widespread algorithm for generating (pseudo) random real numbers that are uniformly distributed between 0 and 1 is the *linear congruential sequence*

$$z_{j+1} = a \cdot z_j + b \pmod{m} \qquad (j = 0, 1, \ldots), \tag{8.16.1}$$

based on four so-called magic numbers, which are *the modul* m $(m > 0)$, *the multiplier* a $(0 \leq a < m)$, *the increment* b $(0 \leq b < m)$, and the *starting value* z_0 $(0 \leq z_0 < m)$. The equality in (8.16.1) is *modulo* m; that is, you take the smallest nonnegative remainder obtained after dividing by m. For example, $7 = 3 \pmod{4}$ and $11 = 2 \pmod{3}$. Division by the modul m yields the desired sequence $x_j = z_j/m$ $(j = 0, 1, 2, \ldots)$ of (pseudo) random numbers $x_0, x_1, \ldots$ satisfying $0 \leq x_j < 1$.

The statistical quality of the sequence $x_0, x_1, \ldots$ depends heavily on a proper selection of m, a, c, and z_0 (see Knuth, 1981, p. 9 + ff.). A statistically sound random number generator produces a sequence $x_0, x_1, \ldots$, such that these values can be regarded as realizations of independent random variables having a uniform distribution in the unit interval.

To simulate the random draw of an r/s lottery, simply generate uniform random numbers $y_1, y_2, y_3, \ldots$ (for example, by repeatedly pressing the button *RND* on your pocket calculator), and put

$$j_n = [s \cdot y_n] + 1, \qquad (n = 1, 2, 3, \ldots)$$

where for a real number y, $[y]$ denotes the largest integer not greater than y. The values $j_1, j_2, j_3, \ldots$, are numbers from the range $1, 2, \ldots, s$. Since these are not necessarily distinct (but your r lotto numbers have to be), discard any repeatedly appearing value among $j_2, j_3, \ldots$ until you have obtained r distinct numbers, which form the desired random (quick-pick) combination. For example, if you want to generate a combination in a 6/49 lottery and your numbers $j_1, j_2, \ldots$ are $15, 43, 8, 21, 43, 7, 11, \ldots$, the only number to discard is 43; thus, the resulting combination will be 15, 43, 8, 21, 7, 11.

References

Bellhouse, D. R. (1991). The Genoese Lottery. *Statistical Science, 6*, 141–148.

Billingsley, P. (1968). *Convergence of Probability Measures.* New York: Wiley.

Blom, G., Holst, L., and Sandell, D. (1994). *Problems and Snapshots from the World of Probability.* New York: Springer.

Bosch, K. (1994). *Lotto und andere Zufälle. Wie man die Gewinnquoten erhöht.* Braunschweig: Vieweg.

Clotfelter, C. T., and Cook, P. J. (1991). *Selling Hope: State Lotteries in America.* Cambridge: Harvard University Press.

Feller, W. (1966). *An Introduction to Probability Theory and Its Applications* (Vol. II). New York: Wiley.

Feller, W. (1970). *An Introduction to Probability Theory and Its Applications* (3rd ed., Vol. I). New York: Wiley.

Gibbons, J. D., and Chakraborti, S. (1992). *Nonparametric Statistical Inference.* New York: Marcel Dekker.

Graybill, F. A. (1976). *Theory and Application of the Linear Model.* Belmont, CA: Wadsworth.

Henze, N. (1982). Verhalten des Chi-Quadrat-Tests für Prüfung der Gleichwahrscheinlichkeit der Lotto-Zahlen bei Nichtgültigkeit der Hypothese. *Metrika, 29*, 135–139.

Henze, N. (1995). The Distribution of Spaces on Lottery Tickets. *Fibonacci Quarterly, 33*, 426–431.

Henze, N. (1997). A Statistical and Probabilistic Analysis of Popular Lottery Tickets. *Statistica Neerlandica, 51*, 155–163.

Henze, N. (1998). "Drawings Since Hit Tables in Lotteries" and a New Multivariate Geometric Distribution. *Statistics & Probability Letters.*

Johnson, N. L., and Kotz, S. (1977). *Urn Models and Their Application.* New York: Wiley.

Joe, H. (1993). Tests of Uniformity for Sets of Lotto Numbers. *Statistics & Probability Letters, 16*, 181–188.

Kadell, D., and Ylvisaker, D. (1991). Lotto Play: The Good, the Fair, and the Truly Awful. *Chance, 4*, 22–25.

Knuth, D. E. (1981). *The Art of Computer Programming* (Vol. II). Reading, MA: Addison-Wesley.

Labat, J.-B. (1730). *Voyages en Espagne et en Italie dans les Années 1705 et 1707* (2). Paris: Delespine.

Lee, J. R. (1972). *Advanced Calculus with Linear Algebra.* New York: Academic Press.

Leite, J. G., de Bragança Pereira, C. A., and Rodrigues, F. W. (1993). Waiting Time to Exhaust Lottery Numbers. *Communications in Statistics—Theory and Methods, 22*, 301–310.

Lottery Player's Magazine. (1994). Vol. 14, No.8. Moorestown, NJ: Regal.

Morgenstern, D. (1979). Der Chi-Quadrat-Test für Prüfung der Gleichwahrscheinlichkeit der Lotto-Zahlen. *Mathematisch-Physikalische Semesterberichte, 26*, 36–39.

Morton, R. H. (1990). Lotto, Then and Now. *The New Zealand Statistician, 25*, 15-21.

Philippou, A. N., and Makri, F. S. (1985). Longest Success Runs and Fibonacci-Type Polynomials. *Fibonacci Quarterly, 23*, 338–346.

Riedwyl, H. (1990). *Zahlenlotto. Wie man mehr gewinnt.* Bern: Paul Haupt.

Stadje, W. (1990). The Collector's Problem with Group Drawings. *Advances in Applied Probability, 22*, 866–882.

Sullivan, G. (1972). *By Chance a Winner: The History of Lotteries.* New York: Dodd, Mead.

Volkmann, J. J. (1771). *Historisch-Kritische Nachrichten von Italien.* Band 3. Leipzig: Fritsch.

Ziemba, W. T. (1986). *Dr. Z's Lotto Guidebook.* Vancouver: Dr. Z Investments.

Index

SAT NOS

2 3 24 27 33 44

4 5 20 26 28 40

6 13 25 34 35 43

7 14 22 23 36 42

8 16 30 31 37 45

12 15 17 18 32 46